KV-555-383

£12.50

016335

Developmental
MANAGEMENT

MANAGING IN THE
INFORMATION SOCIETY

Developmental
MANAGEMENT

Managing in the Information Society

RELEASING SYNERGY JAPANESE STYLE

YONEJI MASUDA

WITH A FOREWORD BY
RONNIE LESSEM

Basil Blackwell

Copyright © 1980 by Institute for the Information Society;
Foreword © Ronnie Lessem 1990

Basil Blackwell Ltd
108 Cowley Road, Oxford, OX4 1JF, UK

Basil Blackwell, Inc.
3 Cambridge Center
Cambridge, Massachusetts 02142, USA

Originally published as *The Information Society as Post-industrial Society*,
© Institute for the Information Society, Tokyo, 1980

This edition first published in 1990

All rights reserved.Except for the quotation of short passages for the purposes of
criticism and review, no part of this publication may be reproduced, stored in a
retrieval system, or transmitted, in any form or by any means, electronic, mechanical,
photocopying, recording or otherwise, without the prior permission of the publisher.

Except in the United States of America, this book is sold subject to the condition that
it shall not, by way of trade or otherwise, be lent, re-sold, hired out, or otherwise
circulated without the publisher's prior consent in any form of binding or cover other
than that in which it is published and without a similar condition including this
condition being imposed on the subsequent purchaser.

British Library Cataloguing in Publication Data

A CIP catalogue record for this book is available
from the British Library.

Library of Congress Cataloging in Publication Data
Masuda, Yoneji,
Managing in the information society/Yoneji Masuda;
with a foreword by Ronnie Lessem.
p. cm. — (Development management)
Rev. ed. of: The information society, 1980.
ISBN 0–631–17575–X.
1. Technology and civilization. 2. Social history – 1970–
3. Information science. 4. Computers and civilization.
5. Information services – Social aspects – Japan. I. Masuda, Yoneji,
Information society. II. Title. III. Series.
HM221.M325 1990
303.48′34 – dc20   89–18382 CIP

Typeset in 11 on 13pt Ehrhardt
by Hope Services (Abingdon) Ltd.
Printed in Great Britain by
William Clowes, Beccles, Suffolk

006

36472

Contents

Preface to the Second Edition

This second edition of my book, '*The Information Society as Post-Industrial Society*' is published by Basil Blackwell of Oxford.

It is about ten years since the first edition of the book was published, and in this period five overseas editions have been published, in Korean, Portuguese, Spanish, Swedish and Hungarian.

It is of course a great pleasure and honour for me that my book has received such a high social evaluation and won so many readers in that wide area of the world.

On the occasion of considering revision of the book, I examined the whole construction and concept of the work, and have reached the conclusion that there is no need to revise what I have written. But it seems appropriate to delete the first two chapters, as these historical descriptions have become obsolete, and to add a concluding chapter to complement the book overall.

As a concluding chapter I have selected 'The Genesis of *Homo Intelligens*'. This is based on my historical hypothesis that the information revolution will bring about not only the transformation of society, but the transformation of man himself, the genesis of the human species. It is hoped that as a result the contents of the book will become deeper and the book itself will prove to be more systematic.

Yoneji Masuda
Japan, 1990

Preface to the First Edition

Mankind is now entering a period of transformation from an industrial society to an information society, and my aim in this book is to attempt a prediction of its character and structure, and to present an overall view of what such a society would be.

When we look back over the development of human society, we see that human history has embraced three types of society: hunting, agricultural and industrial. It is important to note that rapid innovations in the system of societal technology have usually become the axial forces that have brought about these societal transformations. Such a system of societal technology shows four fundamental characteristics:

1 Many different kinds of innovational technology come together to constitute one complex technological system.
2 These integrated systems of technology spread throughout society and gradually become established.
3 The result is a rapid expansion of a new type of productivity.
4 The development of this new type of productivity has societal impact sufficient to bring about the transformation to new societal forms of what had become traditional.

Hunting society proceeded from innovation in societal technology based on systems concerned with hunting. Similarly, the bases for societal transformation first to agricultural society and then to industrial society were innovations in systems of societal technology concerned with agriculture and industrial production. Man is now standing at the threshold of a period of innovation in a new societal technology based on the combination of computer and communications technology. This is a completely new type of societal technology, quite unlike any of the past. Its substance is information, which is invisible.

This new societal technology will bring about a transformation in society which, in a double sense, is unprecedented.

First, the transformation of society is the result of innovations in societal technology, which, in the past, have always been concerned with physical productivity. Even these rapid expansions of physical productivity brought about a change from the feudalistic self-sustained economic system based on agricultural production to a freely competitive goods economy based on industrial production.

Second, the current innovation in societal technology, however, is not concerned with the productivity of material goods, but with information productivity, and for this reason can be expected to bring about fundamental changes in human values, in trends of thought, and in the political and economic structures of society. It will be necessary to build a new paradigm boldly which is free of traditional concepts in order to offer an image of this future information society. This can be done by using the historical analogy and pattern analysis of past societies. Reducing the structure of human society into major components, such as values, trends of thought, innovational technology, the market, economic structure, political systems, I propose to present the pattern of a new concept of each of these components with which to construct an overall picture of the future information society. I will place the major emphasis on a pattern analysis of industrial society, and the historical analogy that applies to the information society.

This book consists of two parts. Part I draws a picture of the overall composition of information society as opposed to industrial society, and makes projections on the realization process of information society, while analysing the developmental stages of actual computerization. To conclude I make a prediction on when information society will be realized, by comparing the tempo of the motive power revolution and that of the computer revolution.

Part II comprises the author's theoretical and conceptual studies on information society. First, I discuss the essential nature of computer-communications technology, the basic characteristic features of the information epoch brought about by this technology, and its social and economic impact. From there I go on to the elements of my conceptual framework for the information society: globalism, time-value, the goal principle, the information utility, a synergetic economic system, information democracy, participatory democracy, voluntary communities, and finally, a vision of 'Computopia'.

This book is a completely rewritten version, for English publication, of the book *Information Economics*, published by Sangyo Noritsu University Press, and parts of the book have already been issued in abbreviated form as listed below.

'Societal Impact of Computerization – Application of the Pattern Model to Industrial Society', Proceedings of the First International Future Research Conference, Kyoto, Japan, April 1970.

'A New Development Stage of the Information Revolution', Applications of Computer/Telecommunications Systems, OECD, November 1972.

'Management of Information Technology for Developing Countries – Adaptation of Japanese Experience to Developing Countries', Data Exchange, April 1974, Diebold Europe.

'The Conceptual Framework of Information Economics', IEEE Transaction on Communications, vol. Com-23, no. 10, October 1975.

'Automated States vs. Computopia: Unavoidable Alternatives for the Information Era', The Next 25 Years, Crisis and Opportunity, World Future Society, 1975.

'Conceptual Structure of Information Economics', Proceedings of the Third International Conference on Computer Communication, Toronto, Canada, August 1976.

'A New Era of Global Information Utility', Proceedings of Eurocomp 78, London, May 1978.

'Future Perspectives for Information Utility', Proceedings of the Fourth International Conference on Computer Communication, Kyoto, Japan, September 1978.

'Privacy in the Future Information Society', Computer Networks, Special Issue, June 1979.

It would be impossible for me to express fully my deep and sincere appreciation to my mentors, family, friends and colleagues in various parts of the world, including Japan.

First of all, I would like to dedicate this book with deep veneration to the memory of the late Reverend Seiichi Yukawa, mentor of long standing, who invoked divine protection for me to complete the publication of this book, but died before it could see the light of day.

Then I would like to thank from the bottom of my heart my wife

Fujie and my son Shigeru, for always showing affection and deep understanding of my work.

I also would like to express my gratitude to Mr Douglas Parkhill and Mr Tomas Ohlin who provided me with valuable government data and made constructive suggestions.

Finally, I would like to thank Bernard Halliwell for his perseverance in the very difficult task of translating the Japanese manuscript into English. His monumental effort to understand the concepts and presentation of ideas in this book has made this English version possible.

I should also wish to make specific mention of the long collaboration of Atsushi Yamada who translated my many articles for the various international conferences. Mrs Maria Radon's proof-reading of the galley sheets is deeply appreciated and reflected her perseverance and understanding. The preparation of the English text and the typing of the MS for this book were the work of Andrew Hughes and his wife Tomoko, for which I would like to express my sincere appreciation.

Figure 1.1 is reproduced by courtesy of the Japan Computer Usage Development Institute, and figures 1.2 and 1.3 by courtesy of the Living Visual Information System Development Association.

Yoneji Masuda
Japan 1980

Foreword: Business in an Information Society

Introduction

The Japanese futurologist Yoneji Masuda was one of the leading contributors to the First Global Conference on the Future in 1980, together with America's Herman Kahn, Italy's Aurelio Peccei and France's Bertrand de Jouvenel. Today he serves as a consultant to numerous government agencies and business corporations, and is a frequent speaker at international conferences around the world.

Masuda is one of the pioneers of computerization in Japan and founder of the Institute for the Information Society. He played a leading role in developing in the 1970s 'The Plan for an Information Society: Japan's National Goal Towards the Year 2000'. Following on its heels were some 20 books including: *Introduction to Management Information Systems*; *Computopia*, his bestseller about a computer-based utopia; and *Managing in the Information Society*.

In this book on the information society, Masuda mixes idealism and realism and Japanese spirit with Western technique. In the process he draws on a Western intellectual tradition that stretches from America to Europe, from Henry Ford to Rudolf Steiner.

Indeed, a line of continuity can be traced between the futurologist Masuda and business innovator Henry Ford, Austrian social visionary Rudolf Steiner, Dutch management thinker Bernard Lievegoed, political philosopher Mary Parker Follett, and social psychologist Soshana Zuboff. But what does Masuda have in common with Ford and the others?

Production as Liberation

In 1926 Henry Ford described the function of the machine to be that of liberating man from brute burdens, so as to 'release his energies to the building of intellectual and spiritual powers for conquests in the fields of thought and higher action'.[1] This is not only a far cry from the scientific management underlying so called 'Fordism', but also an apparent contradiction of the conventionally materialistic wisdom underpinning mass production.

Economics as Fraternity

While Ford was telling Americans of the virtues of 'higher action', Steiner was telling Europeans of the virtues of the division of labour. To provide for oneself, he says, is to work for one's living, 'to work for others is to work out of a sense of social needs'.[2] The division of labour, in that context, is a manifestation of altruism. It is this fraternal nature of business and economics, at its most highly evolved, that Lievegoed, duly following Steiner, has built into his 'developing business'.

Democracy as Association

Mary Parker Follett, a contemporary of Ford and Steiner, was also influential in the 1920s, particularly amongst progressive European business leaders such as Seebohm Rowntree in Britain and Walter Rathenau in Germany. Parker Follett took a fresh look at democracy. For her democracy was not a spreading out, this was merely its external aspect, but a drawing together: 'We have an instinct for democracy because we have an instinct for wholeness'.[3] The twentieth century, she maintained, fraught by economic and political warfare, needed to find a new principle of association. However, Parker Follett would have to wait another 60 years for people and technology to catch up with her argument.

The 'Informating' Environment

The contemporary social psychologist Soshana Zuboff, in many ways a latter day Parker Follett, is less sanguine than Ford or Steiner are about the benefits of the division of labour. For her, such scientific management hardens the boundaries between man and man, between one social class and another. However, at the same time it did set in train an evolutionary process whereby knowledge played an increasingly important part in management. This process reaches its climax, through computerization, as a process of 'informating' sets in.

According to Zuboff, by redefining the grounds of knowledge from which competent behaviour is derived, new information technology lifts skill from its historic dependence upon a labouring, sentient body. Such fundamental adaptations involve the quality of mind that a knowledge worker brings to the computer screen, 'the capacity to paint an inward picture of the world to which the data belong. This *inner vision*, constructed out of a combination of memory and imagination, must take the place of a world that is ready-at-hand'.[4] [Emphasis added.]

The Information Society

From Zuboff's micro perspective, focused on the mind of the individual knowledge worker, we turn to Masuda's macro perspective, oriented towards a future information society as a whole: 'If the goal of industrial society is represented by volume consumption of durable consumer goods or realization of heavy mass consumption, information society may be termed a society with highly intellectual creativity where people may draw future designs on an invisible canvas and pursue and realize individual lives worth living'.[5]

Remarkably, Masuda combines a Japanese orientation towards the interdependent whole – 'the information society will embody the principle of synergy and social benefit'[6] – with a European or American orientation towards the self-actualizing individual. In that respect he is truly a world figure. Let us delve a little further, then, into the American and European heritage upon which he draws before returning to his own Eastern perspective.

Ford on Management

At face value, Henry Ford, as the great orchestrator of mass production and large-scale industrialization, would seem to represent the very antithesis of Yoneji Masuda. Yet paradoxically there is much in Ford's philosophy of industry that anticipates Masuda.

The Function of the Machine is Liberation

Ford was a man before his time. He dreamt not only of creating a means of powered transportation for every man, but also of releasing him from the physical burdens of life, so that he could develop his emotional and spiritual capacities. The highest use of capital, he said, is not to make more money, but to make money do more service for the betterment of life.

In the pursuit of his dream, Ford was somewhat blind to the alienating effects of his assembly line technology. On the other hand, we can now see, with the benefit of hindsight, that mechanization and mass production were necessary forerunners of the information age. We could not have leapt straight from small-scale, cottage industry to large-scale automation. The seeds that Ford sowed, despite his alienating technology and division of labour, were those of partnership between worker and manager. He thus anticipated the synergistic approach that Masuda was to adopt some 60 years later.

'The Economic Fundamental is Labour'

When Ford first wrote that 'the economic fundamental is labour'[7] rather than land, capital or even entrepreneurship, management may have been forgiven for ignoring his words of wisdom. After all, it is difficult to envisage the eminently replaceable man on the assembly line as the heart of the business, outside of a humanitarian perspective. Yet in today's information society the notion of the 'knowledge worker' as the economic fundamental is no longer, as indicated in a recent book on 'Managing Knowhow', so far-fetched.

Human Being: the Knowhow Machine

The human being is the 'machine' of the knowhow company. The greater the number of able people there are in an organization the greater its productive capacity.

The human being is the only productive resource in such companies. The journalist, the research engineer, the programmer, the consultant and the physician, the knowhow professionals, all work by exploiting knowhow. If the doctors fall ill there are no robots to replace them.[8]

Though reference to human beings as machines is of dubious worth, the general point that Tom Lloyd, an English financial journalist, and Karl Erik Sveiby, a Swedish management consultant, are making, is an interesting one. Ford, the mechanical engineer, you might say, has been hoist by his own philosophical petard. Like all great innovators, Ford sowed the seeds of his own destruction. Mass and undifferentiated labour, in becoming the economic fundamental, has turned into the individually differentiated knowledge worker. Hence, for Masuda, emphasis shifts from mass production and consumption, centred upon motorization, to highly intellectual and individualized creativity, centred upon computerization.

Steiner on Economics

For Steiner, the division of labour, in concept if not in application, was something different from either Adam Smith's or Henry Ford's version: to provide for oneself is to work for one's living, but to work for others is to work out of a sense of social needs. So far Ford and Steiner are at one. However, Steiner adds: to the extent that a division of labour is present, so might be altruism.

The difference, then, between an altruistic division of labour and a selfish one lies not in the external circumstances but in the mind of the manager or technologist. But to what extent has the labour been divided up to yield greater service, or to what extent has mechanization ensued merely to line the pockets of the entrepreneur in question? The fact of the matter does remain, however, at least as far as Steiner is concerned, that mutual interest rather than self interest lies at the heart of economics.

The Economics of Mutual Interest

In Steiner's so-called 'threefold commonwealth', liberty is acquired through the cultural realm, equality through the political sphere and fraternity through economics. In that sense Steiner has more in common with the Japanese than with his fellow Europeans. However, at the same time he is at one with the spirit of individualism. Institutions themselves cannot work socially, he says, for they require socially attuned human beings. Indeed, as an Austrian, he bordered eastern-European collectivism and western-European individualism.

For Steiner, the preservation of individual initiative depends not on Adam Smith's self-serving invisible hand, but on free-forming associations. In that sense he, like Ford, anticipates Masuda. 'Where there is simply a market relationship – where supply and demand are the determining factors – only the egoistic type of value can be considered. The market relationship must be superseded by associations that regulate the exchange and production of goods through an intelligent consideration of human needs. Such associations can replace mere supply and demand by contracts between groups of producers and consumers, and between different groups of producers'.[9]

With the proliferation of joint ventures today, Steiner's words are coming to fruition, just like those of Mary Parker Follett.

Follett on the New State

For Mary Parker Follett freedom lay in the harmonious and unimpeded working of the laws of one's own nature, and the true nature of every man is found only in the whole. A man is ideally free only so far as he is interpenetrated by every other human being. In other words, he gains his freedom through a complete set of relationships.

The Principle of Association

As for the individual, so for the organization and for the whole of society. The twentieth century, according to Parker Follett, must find a new principle of association. Crowd philosophy and national patriotism, just like a parochial corporate mentality, must go. Group organization, rather, is to be the new method in politics, the basis for the future industrial system and the foundation of international order.

Life is enriched, Parker Follett says, by collaboration with all the powers of the universe. Man lives on several planes – psychological, economic and political – and his development depends on the uniting of these. Her definition of individuality, for the company or for the person, is 'finding my place in the whole'.[10] The business world, therefore, is never again to be directed by individual intelligences, but by intelligences interacting and ceaselessly influencing each other.

Zuboff in the Age of 'the Smart Machine'

Soshana Zuboff, at the Harvard Business School, picks up where her illustrious predecessors leave off. As a representative of the information age, she has a particular interest in the influence of what she calls an 'informating' technology on our organizational futures.

Automating versus Informating Technology

By redefining the grounds of knowledge from which competent behaviour is derived, Zuboff claims, the new information technology lifts skill from its historic dependence upon the physical body. In other words, the traditional notion of craftsmanship, dependent upon physical touch and feel, is displaced. The nature of this displacement, however, is variable, depending upon whether it serves a mere 'automating' or also an 'informating' function.

As long as the technology is treated narrowly in its automating function, Zuboff says, it perpetuates the logic of the industrial machine that, over the course of this century, has made it possible to rationalize work while decreasing the dependence on human skills. However, 'when the technology also informates the processes to which it is applied, it increases the explicit information content of tasks and sets into motion a series of dynamics that will ultimately reconfigure the nature of work and the social relationships that organize productive activity'.[11]

Whereas automation, according to Zuboff, treats as negligible the potential value to be added from human learning, the *informating process* takes *learning* as its *pivotal experience*. Its objective is to achieve the value that can be added from learning. The informating manager in fact assumes that, through making the organization more transparent, valuable communal insight will be gained.

Cedar Bluff Pulping Manager

Once operators had established the referential function of the data, many moved to a higher level of complexity in dealing with the system of electronic symbols. At this level, the problem was not only to clarify the significance of individual data elements but also to construct from these elements, and particularly from their combinations, an interpretation of the abstract properties of the production process.

Instead of a problem of correspondence, the data now presented an opportunity for insight into functional relationships, states, conditions, trends, likely developments, and underlying causes, none of which can be reduced to a concrete, external referent. In other words, data needed to be translated into information, and information into insight.[12]

Drivers of People: Drivers of Learning

In an informated environment, then, the electronic text displays the organization's work in a new way. Much of the information and knowhow that was private becomes public. Personal gain depends less upon maintaining private knowledge than upon developing mastery in the interpretation and utilization of the public, dynamic and electronic text.

'Communicative competence', Zuboff tells us, 'requires psychological individuation, which introduces a new sense of mutuality and equality into group life. Hierarchical or other status based distinctions hold less power in a group of individuated and competent interpreters, each with access to the metalanguages of choice and innovation'.[13] Whereas the worker's knowledge had been implicit in action, the informating process makes the knowledge explicit.

Learning increases the pace of change. For an organization to pursue an informating strategy it must maximize its own ability to learn and explore the implications of that learning for its long-range plans about markets, product development and new sources of competitive advantage. Some members will need to guide and coordinate learning efforts in order to lead an assessment of strategic alternatives and to focus organizational intelligence in areas of strategic value. The increased time horizon of these managers' responsibilities provides the reflective distance with which they can gauge the quality of the learning environment and can guide change that will improve collective learning.

However, we remain, Zuboff says, in the final years of the twentieth century, prisoners of a vocabulary in which 'managers' require 'employees'; 'superiors' have 'subordinates'; jobs are defined to be 'specific', 'detailed', 'narrow' and 'task related'; and organizations have levels that in turn make possible 'chains of command' and 'spans of control'. The guiding metaphors are, in fact, military. Therefore, a new division of learning requires a new vocabulary, that is one of 'colleagues' and 'co-learners', of 'exploration', 'experimentation' and 'innovation'. Jobs in this new learning environment are comprehensive; tasks are abstractions that depend on insight and synthesis; and power is a roving force that comes to rest as dictated by function and need. The contemporary language of work, Zuboff and Masuda would agree, are inadequate to express these new realities.

The New Bankers – Global Bank

Their vision was one of a group of people with various competencies brought together around the data base in order to collaborate in the construction of meaning that would lead to the identification of opportunities for innovation. One manager described the data base as the new 'vault' that contained the bank's real assets. Intellective mastery and teamwork would provide the keys to the vault.[14]

The relationships that characterize a learning environment can thus be thought of as post-hierarchical. This does not imply that differentials of knowledge, responsibility and power no longer exist; rather, they can no longer be assumed. Instead they adapt and develop their character in relation to the situation, the task and the actors at hand. Managing such intricate relationships, Zuboff maintains, calls for a new level of action-centred skill, placing a high premium on intuitive understanding in the course of acting with people. Moreover, it is the dictates of a learning environment rather than those of command structure which now shape the development of such interpersonal knowhow.

Concentric Circles

As the intellective skill base becomes the organization's most precious resource, managerial roles function in a particular way to enhance its quality. Organizational members can be thought of as being arrayed in concentric circles around a central core, which is the electronic data base. Because intellective skill is relevant to the work of each ring of responsibility, the skills of those who manage daily operations form an appropriate basis for their progression into roles with more comprehensive responsibilities.

The vision of a concentric organization is one that seems to rely on metaphors of *wholeness*, *interdependency* and *integration*. Zuboff asks: 'What is required of managers in such a workplace, where learning and integration constitute the two most vital priorities?' and 'How is

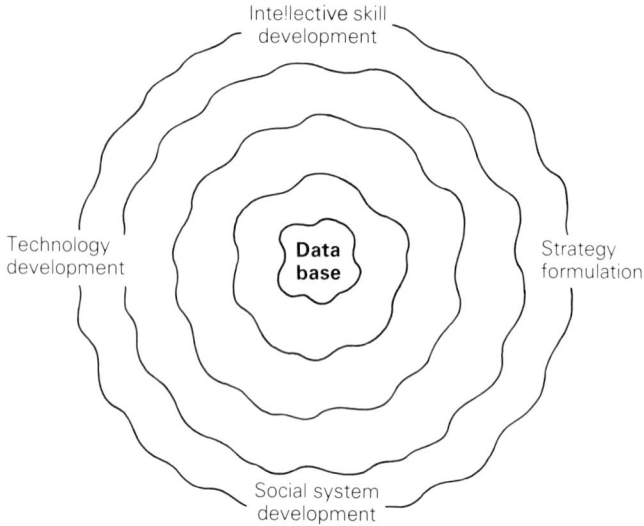

Figure 1 Concentric Responsibility.

the social system of such an organization to be managed?' In fact, new sources of personal influence are required, ones which are associated with the ability to learn and to engender learning in others, in contrast to an earlier emphasis upon contractual relationships or on the authority derived from function and position. Moreover, the inter-penetration between concentric rings of responsibility provides a key source of organizational integration.

The activities arrayed on the responsibility rings at a greater distance from the core incorporate at least these four kinds of managerial activity: intellective skill development, technology development, strategy formulation and social system development. This means that some organizational members will be involved in both higher order analysis and conceptualization as well as in promoting learning and skill development among those with operational responsibility. Their aim is to expand the knowledge base and to improve the effectiveness with which data are assimilated, interpreted and responded to. They have a central role in creating an organizational environment that initiates learning and in supporting those in other managerial activities to develop their talents as educators and learners.

Zuboff's analysis leads us right into Masuda's information society.

For in citing the computer as the innovative technology within such a society, he maintains that its fundamental function will be to substitute and amplify the mental labour of man,[15] thus enabling people to pursue lives worth living.

Masuda's Image of the Future: A Japanese Perspective

Computopia

For Masuda, the so-called 'information society' will be completely different from the one we know today. The grounds for his assertion lie in his belief that the production of 'information values' and not material values will be the driving force behind social and economic change.[16] More specifically, *computer technology* will not only substitute and amplify the mental labours of man but the mass production of *systematized knowledge* will ensue; the *information utility*, consisting of information networks and data banks, will replace the factory as the societal symbol, and become the production and distribution centre for information goods; the *knowledge* frontiers of tomorrow, as opposed to the physical frontiers of yester-year, will yield the potential markets of the future;[17] and the leading industries will be *intellectual industries*. Moreover, according to Masuda, the economy will change from an exchange-based one to a *synergistic economy*; the invisible hand of *laissez-faire* economics will be replaced by the *goal principle*; the *voluntary community* will replace public and private enterprise at the leading edge of social and economic development;[18] the principle of *synergy* and *social benefit* will take the place of profit maximization as the predominating ethic; a *horizontal* and multi-centred society will supplant centralized power and hierarchical classes; the realization of *time-value*, that is, self-fulfilment, for each human being will replace the contemporary welfare state.[19]

Computers: the Core Technology

Objectifying Knowledge

The core technology of the information society is, of course, the computer. As such it has three leading characteristics. The first is the complete *objectification* of knowledge. The process of objectification may be likened to mankind's overall progress in knowledge.[20] The invention of written characters made the preservation of information in objective form portable and economic for the first time in human history.[21] The development of printing, and subsequently photocopying, came second, bringing in their wake the mass production and distribution of knowledge. The computer, thirdly, has made it possible to produce original information by machine, thus amplifying man's mental capacity.[22]

Producing Cognitive Information

The second characteristic is the production of *sophisticated cognitive information*. Such information refers to a projection of the future. Thus it can be used for problem solving and for forecasting.[23] For Masuda it is such cognitive or intellectually centred information, as opposed to the 'affective' or emotionally centred variety, which makes possible active interaction with the external environment.[24]

Forming Information Networks

The third characteristic of computing, arising out of its association with telecommuting, is the *information network*. There are two varieties for Masuda, environmental and organismic networks. 'Environmental information networks' are concerned with the relationship between an organism and the external world.[25] The system of 'organismic information' is concerned with the carrying out of the functions of the organization itself.[26] The combination of the two will lead, according to Masuda, to a *multi-centred complex society*, in which many systems are linked and integrated by complex networks. Such a society will be capable of responding rapidly to change.[27]

Globalism: the Spirit of the Times

The new spirit of the times, within the information society, will be 'globalism', the chief aim of which, for Masuda, will be the liberation of the human spirit. The first characteristic of such globalism will be so-called *spaceship thought*. To the extent that the Renaissance was an era of explosion, this new era, a 'neo-Renaissance', will be one of implosion. The old territorial frontiers that divided man are breaking down, now evident in the Northern hemisphere by Europe 1992.[28]

The second of Masuda's characteristics of globalism is *the idea of symbiosis*. Widespread pollution and evidence of damage to the environment led to the rise of the new science of ecology, which bases its research on symbiosis of man and nature.[29] The recently growing awareness of the dramatic and world-wide implications of 'the greenhouse effect' is a typical case in point.

The third characteristic is the concept of *global information space*. This means space connected by information networks, straddling regional and national boundaries. Information will have no boundaries.[30] As mutual exchanges of information expand, Masuda maintains, so will mutual understanding, touching issues that lie outside the boundaries of nation states. When this happens, the spirit of globalism, prevailing over conflicting national interests and differences, will become broadly and deeply rooted in the minds of the people.[31]

The Information Utility: Symbol of the Information Society

The modern factory, the present societal symbol, supplanted the farm, which had been the production centre of agricultural societies. The information utility, correspondingly, will become the centre of production in industrialized societies of the twenty-first century.[32]

Information Infrastructure

The first fundamental of the information utility, as Masuda conceives of it, is that it will take the form of an *information infrastructure*. As such its central facilities will be equipped with large-scale computers capable of simultaneous parallel processing, connected with large-capacity memory devices, a large number of programme packages and extensive data bases.[33] Such facilities will be provided for the use of the general public, for which purposes the central facilities will be networked out to desktop computers in businesses, schools and homes.[34]

The unique character of this infrastructure, moreover, is founded upon the self-multiplying nature of information. For information, unlike physical commodities, is not consumable – goods are consumed when used, but information remains, however much it is used; it is non-transferable – in the transfer of goods from one place to another they are physically moved from the original source, but in the transfer of information both giver and receiver are able to access it; it is indivisible – materials such as electricity and water are divided for use, but information can only be used as a set; finally, it is accumulative – whereas the accumulation of goods is by their non use, the quality of information is raised by adding new information to what has already been accumulated.[35]

Synergetic Production and Shared Utilization

The second fundamental of Masuda's information utility is that both the *production* and the *utilization* will be a combined operation. This means that the infrastructure of information is self-multiplying. All information is cumulative.[36]

To begin, information will be offered as a public service, whether, for example, on employment opportunities, leisure facilities, or financial offerings. Drawing on these initial services the user will subsequently produce more information for him or herself, making use of appropriately customized software.[37] In addition to this, such information can be shared with third parties, at the discretion of the user. Finally, the synergetic production of information by a group will be made possible through interactive systems.[38]

Citizen Participation

The third fundamental, for Masuda, is citizen participation.[39] He argues that it is through such participation that self-multiplication of information will be maximized. Moreover, it is in this way that the dangerous tendency towards a centralized, bureaucratically controlled society will be prevented.[40]

A Synergistic Economy: the Information Axis

Synergetic Production and Shared Utilization

The onset of Masuda's 'synergetic' economy will be spearheaded first of all by a shift from a free enterprise economy to a synergistic production and shared utilization of the information infrastructure. In other words, not only will the information utilities be used by the people themselves to produce the information they require, but the data they collect will be available for shared utilization with others.[41] Since producers will be users, the information utilities will be axial institutions in the economies of the information societies and the joint production and shared utilization of information goods will greatly influence the economic structure as a whole.[42]

Voluntary Synergy to Achieve a Shared Economic Goal

The second aspect of Masuda's economic axis will be, in contrast with the previous era of free enterprise, a voluntary synergy. Because we may have begun to realize that we live on spaceship earth, we need to give priority to economic activities that correspond with shared goals.[43]

Exercising Self-Restraint

Masuda, coming from a society that has made self-discipline and self-restraint its hallmark, argues, thirdly, that autonomous constraints on the consumption of goods will need to apply to ensure the stabilized development of the global economy. Current levels of budget deficit, in industrialized societies like Britain and America, not to mention some of the countries in the Third World, may serve to reinforce Masuda's argument.[44]

Developing Functional Synergy

The tendency for both the workforce and the general public to participate in the ownership of enterprise has become common practice; the information society will, fourthly, serve to accelerate this trend.[45]

The relationships in the traditional free enterprise economy, Masuda argues, are based on authority exercised from above on those below. But information-based enterprises of the future will move towards an economic community of people who participate voluntarily and share the same goal. The synergetic relations that come into being will not be authoritarian but purely functional.[46]

Participatory Democracy: Nexus of Policy-making

The Call for Participatory Democracy

Masuda gives four reasons why the call for participatory democracy has become both urgent and appropriate.

1 The behavioural pattern of ordinary citizens, particularly in industrialized societies, is changing. People are becoming relatively less concerned with material wants and more concerned with self-fulfilment.[47]

2 The powers of public and private enterprises have expanded so massively that there is a need for ordinary citizens to have countervailing power.[48]

3 Many of the questions that we have to decide are matters that concern the whole of mankind. They are global issues that know no national boundaries and the settlement of which directly affects the lives of ordinary persons.[49]

4 Technical difficulties that until now had made it impossible for large numbers of citizens to participate in policy making have now been solved by the revolution in computer-communications technology.[50]

The Principles of Participative Democracy

Masuda sets out six basic principles of participative democracy within an information society.

1 All citizens would have to participate in decision-making.[51]

2 The spirit of synergy and mutual assistance would have to permeate the whole system.[52]

3 All relevant information should be available to the public.[53]

4 All benefits received and sacrifices made by citizens should be distributed equitably amongst them.[54]

5 A solution should be sought by agreement.[55]

6 Once decided, all citizens would be expected to cooperate in applying the solution.[56]

Voluntary Communities: Core of the Social Structure

Neither the public nor the private sector but the voluntary community lies at the core of Masuda's information society.[57] In such communities ties to a physical locality will not be as strong as in the past because each will occupy information space, that is, shared ideas and goals.[58] 'Cooperative labour' will replace the more conventional division of labour. In fact, there will be a growth of multi-centred, multi-layered communities, open to the outside world. Each community, while maintaining its independence, will be interlinked via communication networks to others. The business sector, moreover, will establish mutually dependent relationships with these voluntary communities.[59]

Computopia: Rebirth of 'Theological Synergism'

Finally, Masuda shares with us his vision of a 'Computopia'. Extending Adam Smith's original idea of a 'universally opulent society', Masuda arrives at six features of 'theological synergism'.[60]

1 In his vision of 'computopia', each individual will pursue and realize time-value.[61]
2 There will be freedom of decision and equality of opportunity. In other words, people will share equally in the opportunity to realize time-value.[62]
3 There will be a flourishing of diverse voluntary communities. The development of information productive power will liberate man from subsistence labour.[63]
4 There will develop interdependent synergistic societies. A synergistic society is one where individuals and groups cooperate to achieve the common goals set by society as a whole.[64]
5 There will be the realization of functional societies free of overruling power. A synergistic society is one where the independence of the individual harmonizes with the order of the group.[65]
6 The goal of 'Computopia' is the rebirth of theological synergism of man and the Supreme Being. Spiritual rebirth, says Masuda, depends on the cooperation of the will of man and the grace of God.[66]

RONNIE LESSEM
London, 1990

Notes

1 H. Ford, *My Life and Work*, Doubleday, New York, 1922, p. 167.
2 R. Steiner, *The Social Future*, Rudolf Steiner Press, London, 1985.
3 M. Parker Follett, *The New State*.
4 S. Zuboff, *In the Age of the Smart Machine*, Heinemann, London, 1988, p. 116.
5 Y. Masuda, *Managing in the Information Society*, Basil Blackwell, Oxford, 1990, pp. 131–2.
6 Ibid., p. 9.

xxxii   FOREWORD BY RONNIE LESSEM

7 H. Ford, *My Life and Work*, p. 36.
8 T. Lloyd and K. Sveiby, *Managing Knowhow*, Bloomsbury Press, London, 1988, p. 11.
9 R. Steiner, *The Social Future*, p. 41.
10 M. Parker Follett, *The New State*, p. 112.
11 S. Zuboff, *In the Age of the Smart Machine*, p. 11.
12 Ibid., p. 92.
13 Ibid., p. 206.
14 Ibid., p. 202.
15 Y. Masuda, *Managing in the Informations Society*, p. 3.
16 Ibid., p. 3.
17 Ibid., p. 4.
18 Ibid., p. 5.
19 Ibid., p. 9.
20 Ibid., p. 26.
21 Ibid., p. 26.
22 Ibid., p. 28.
23 Ibid., p. 28.
24 Ibid., p. 29.
25 Ibid., p. 32.
26 Ibid., p. 33.
27 Ibid., p. 35.
28 Ibid., p. 46.
29 Ibid., p. 47.
30 Ibid., p. 47.
31 Ibid., p. 48.
32 Ibid., p. 53.
33 Ibid., p. 53.
34 Ibid., p. 54.
35 Ibid., p. 55.
36 Ibid., p. 56.
37 Ibid., p. 56.
38 Ibid., p. 57.
39 Ibid., p. 57.
40 Ibid., p. 60.
41 Ibid., p. 77.
42 Ibid., p. 78.
43 Ibid., p. 79.
44 Ibid., p. 79.
45 Ibid., p. 79.
46 Ibid., p. 79.
47 Ibid., p. 81.
48 Ibid., p. 83.
49 Ibid., p. 83.
50 Ibid., p. 83.

51 Ibid., p. 84.
52 Ibid., p. 84.
53 Ibid., p. 85.
54 Ibid., p. 86.
55 Ibid., p. 86.
56 Ibid., p. 87.
57 Ibid., p. 120.
58 Ibid., p. 124.
59 Ibid., p. 125.
60 Ibid., pp. 130–1.
61 Ibid., p. 132.
62 Ibid., p. 133.
63 Ibid., p. 134.
64 Ibid., p. 135.
65 Ibid., p. 135.
66 Ibid., pp. 139–40.

PART I

Emergence of the
Information Society

MATTHEW BOULTON
TECHNICAL COLLEGE
COLLEGE LIBRARY

I

Image of the Future Information Society

What is the image of the information society? The concept will be built on the following two premises:

1 The information society will be a new type of human society, completely different from the present industrial society. Unlike the vague term 'post-industrial society', the term 'information society' as used here will describe in concrete terms the characteristics and the structure of this future society. The basis for this assertion is that *the production of information values and not material values will be the driving force* behind the formation and development of society. Past systems of innovational technology have always been concerned with material productive power, but the future information society must be built within a completely new framework, with a thorough analysis of the system of computer-communications technology that determines the fundamental nature of the information society.
2 The developmental pattern of industrial society is the societal model from which we can predict the overall composition of the information society. Here is another bold 'historical hypothesis': *the past developmental pattern of human society can be used as a historical analogical model for future society.*

Putting the components of the information society together piece by piece by using this historical analogy is an extremely effective way of building the fundamental framework of the information society.

The Overall Composition of the Information Society

Table 1.1 displays the overall framework of the information society based upon these two premises. It presents the overall composition of the information society based on a historical analogy from industrial society. Let me explain each of the major items. Of course the entire picture of the future information society cannot be given at this stage, but at least table 1.1 will help the reader understand the composition of and overall relations between chapters that unfold later in the book.[1]

1 The prime innovative technology at the core of development in industrial society was the steam engine, and its major function was to substitute for and amplify the physical labour of man. In the information society, 'computer technology' will be the innovational technology that will constitute the developmental core, and its fundamental function will be to *substitute and amplify the mental labour of man.*

2 In industrial society, the motive power revolution resulting from the invention of the steam engine rapidly increased material productive power, and made possible the mass production of goods and services and the rapid transportation of goods. In the information society, 'an information revolution' resulting from development of the computer will rapidly expand information productive power, and make possible *the mass production of cognitive, systematized information, technology and knowledge.*

3 In industrial society, the modern factory, consisting of machines and equipment, became the societal symbol and was the production centre for goods. In the information society *the information utility* (a computer-based public infrastructure), consisting of information networks and data banks, will replace the factory as *the societal symbol,* and become the production and distribution centre for information goods.

4 Markets in industrial society expanded as a result of the discovery of new continents and the acquisition of colonies. The increase in consumption purchasing power was the main factor in expansion of the market. In the information society, 'the knowledge frontier' *will become the potential market,* and the increase in the possibilities of problem solving and the development of opportunities in a society that is constantly and dynamically developing will be the

primary factor behind the expansion of the information market.

5 In industrial society, the leading industries in economic development are machinery and chemicals, and the total structure
comprises primary, secondary and tertiary industries. In the
information society the leading industries will be *the intellectual
industries*, the core of which will be the knowledge industries.
Information-related industries will be added as *the quaternary group*
to the industrial structure of primary, secondary and tertiary. This
structure will consist of a matrix of information-related industries
on the vertical axis, and health, housing and similar industries on
the horizontal axis.

6 The economic structure of industrial society is characterized by
(1) a sales-oriented commodity economy, (2) specialization of
production utilizing divisions of labour, (3) complete division of
production and consumption between enterprise and household.
In the information society, (1) information, the axis of socio-
economic development, will be produced by the information
utility, (2) self-production of information by users will increase;
information will accumulate, (3) this accumulated information
will expand through synergetic production and shared utilization
and (4) the economy will change structurally from an exchange
economy to *a synergetic economy*.

7 In industrial society the law of price, the universal socio-economic
principle, is the Invisible Hand that maintains the equilibrium of
supply and demand, and the economy and society as a whole
develop within this economic order. In the information society *the
goal principle* (a goal and means principle) will be the fundamental
principle of society, and the synergetic feedforward, which
apportions functions in order to achieve a common goal, will work
to maintain the order of society.

8 In industrial society, the most important subject of social activity
is the enterprise, the economic group. There are three areas:
private enterprise, public enterprise and a third sector of
government ownership and private management. In the information society the most important subject of social activity will be *the
voluntary community*, a socio-economic group that can be broadly
divided into local communities and informational communities.

9 In industrial society the socio-economic system is a system of
private enterprise characterized by private ownership of capital,
free competition and the maximization of profits. In the informa-

Table 1.1 Pattern comparison of industrial society and the information society

		Industrial society	Information society
Innovational technology	Core	Steam engine (power)	Computer (memory, computation, control)
	Basic function	Replacement, amplification of physical labour	Replacement, amplification of mental labour
	Productive power	Material productive power (increase in per capita production)	Information productive power (increase in optimal action-selection capabilities)
Socio-economic structure	Products	Useful goods and services	Information, technology, knowledge
	Production centre	Modern factory (machinery, equipment)	Information utility (information networks, data banks)
	Market	New world, colonies, consumer purchasing power	Increase in knowledge frontiers, information space
	Leading industries	Manufacturing industries (machinery industry, chemical industry)	Intellectual industries (information industry, knowledge industry)
	Industrial structure	Primary, secondary, tertiary industries	Matrix industrial structure (primary, secondary, tertiary, quaternary/systems industries)
	Economic structure	Commodity economy (division of labour, separation of production and consumption)	Synergetic economy (joint production and shared utilization)

Socio-economic principle	Law of price (equilibrium of supply and demand)	Law of goals (principle of synergetic feedforward)
Socio-economic subject	Enterprise (private enterprise, public enterprise, third sector)	Voluntary communities (local and informational communities)
Socio-economic system	Private ownership of capital, free competition, profit maximization	Infrastructure, principle of synergy, precedence of social benefit
Form of society	Class society (centralized power, classes, control)	Functional society (multi-centre, function, autonomy)
National goal	GNW (gross national welfare)	GNS (gross national satisfaction)
Form of government	Parliamentary democracy	Participatory democracy
Force of social change	Labour movements, strikes	Citizens' movements, litigation
Social problems	Unemployment, war, fascism	Future shock, terror, invasion of privacy
Most advanced stage	High mass consumption	High mass knowledge creation
Values	Value standards	Material values (satisfaction of physiological needs)	Time-value (satisfaction of goal achievement needs)
	Ethical standards	Fundamental human rights, humanity	Self-discipline, social contribution
	Spirit of the times	Renaissance (human liberation)	Globalism (symbiosis of man and nature)

INDUSTRIAL SOCIETY

INFORMATION SOCIETY

Triple concept of human society

RENAISSANCE SPIRIT — Humanism material value

INDUSTRIAL REVOLUTION — Steam engine & machinery

Free competition of private enterprises / Profit pursuit

Parliamentary democracy / Labour movement

Modern factory — Natural resources / Useful goods

Mass production → Mass consumption

Industrial wastes / Consumption wastes

Consumption needs
Individualism — Industrial society

Material productive power

Achievement needs
Synergism
Information productive power — Information society

Pollution / Environmental disruption / Exhaustion of natural resources

Freedom from pollution / Symbiosis with nature / Resource-saving

INFORMATION REVOLUTION — Computer & communication network

GLOBALISM — Symbiosis Time-value

Information utility / Data → Information

Joint production / Shared utilization

Synergetic economy / Social contribution

Participatory democracy / Citizens' movement

Global information network / Technology assessment

HIGH MASS CONSUMPTION SOCIETY — Urbanization Motorization Recreation

HIGH WELFARE SOCIETY — Consumerism Social welfare Leisure

HIGH MASS KNOWLEDGE CREATION SOCIETY — Computerization Voluntary community Self-actualization

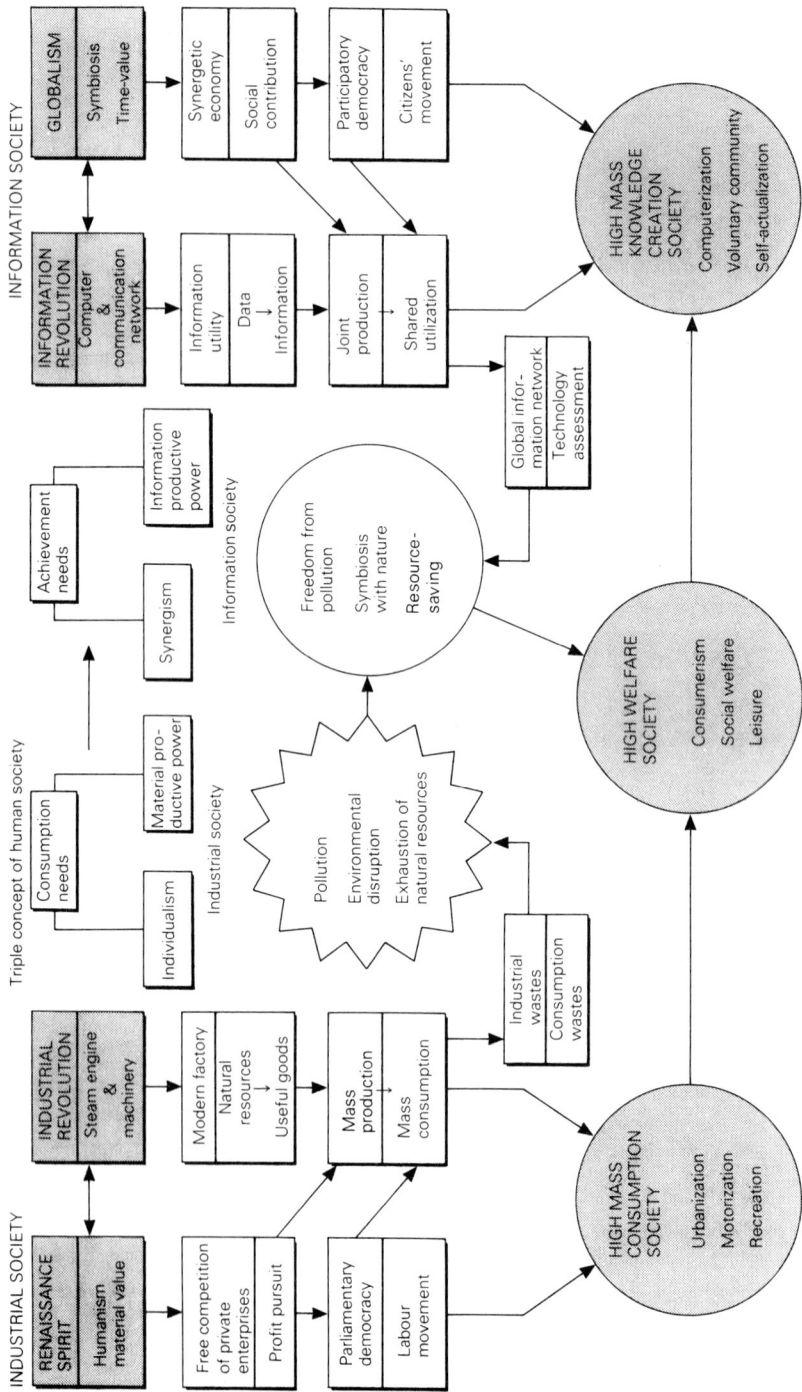

Figure 1.1 Transformation Process from Industrial Society to Information Society.

tion society, the socio-economic system will be a voluntary civil society characterized by the superiority of its infrastructure, as a type of both public capital and knowledge-oriented human capital, and by a fundamental framework that embodies *the principle of synergy and social benefit*.

10 Industrial society is a society of centralized power and hierarchical classes. The information society, however, will be a multi-centred and complementary voluntary society. It will be horizontally functional, maintaining social order by *autonomous and complementary functions of a voluntary civil society*.

11 The goal of industrial society is to establish a Gross National Welfare Society, aiming to become a cradle-to-grave high welfare society. The information society will aim for *the realization of time-value* (value that designs and actualizes future time), for each human being. The goal of society will be for everyone to enjoy a worthwhile life in the pursuit of greater future possibilities.

12 The political system of industrial society is a parliamentary system and majority rule. In the information society the political system will become a *participatory democracy*. It will be the politics of participation by citizens; the politics of autonomous management by citizens, based on agreement, participation and synergy that take in the opinions of minorities.

13 In industrial society, labour unions exist as a force for social change, and labour movements expand by the use of labour disputes as their weapon. In the information society, *citizen movements* will be the force behind the social change; their weapons will be litigation and participatory movements.

14 In industrial society there are three main types of social problems: recession-induced unemployment, wars resulting from international conflict, and the dictatorships of fascism. The problems of the information society will be future shocks caused by the inability of people to respond smoothly to rapid societal transformation, acts of individual and group terrorists such as hijackings, *invasions of individual privacy* and the crisis of *a controlled society*.

15 The most advanced stage of industrial society is a high mass consumption stage, centring on durable goods, as evidenced by motorization (the diffusion of the automobile). The most advanced stage of the information society will be *the high mass knowledge creation society*, in which computerization will make it

possible for each person to create knowledge and to go on to self-fulfilment.

16 In industrial society, the materialistic values of satisfying physiological and physical needs are the universal standards of social values; but in the information society, seeking *the satisfaction of achieved goals* will become the universal standard of values.

17 Finally, the spirit of industrial society has been the Renaissance spirit of human liberation, which ethically means respect for fundamental human rights and emphasis on the dignity of the individual, and a spirit of brotherly love to rectify inequalities. The spirit of the information society will be *the spirit of globalism, a symbiosis in which man and nature* can live together in harmony, consisting ethically of *strict self-discipline and social contribution*.

Notes

1 Y. Masuda, 'Social Impact of Computerization. An Application of the Pattern Model for Industrial Society', *Challenges from the Future*, Tokyo, Kodansha, 1970.

2

When will the Information Society be Realized?

In considering the question of when the information society will become a reality, we first have to look at the stages of development in computerization, utilization and popularization of computers in society, and try to decide what the current stage of development is, and from this to predict when computerization will reach its final stage.

The Four Developmental Stages of Computerization

Table 2.1 breaks down the development of computerization according to the areas affected. There are four stages based respectively on the use of computers at the level of (1) big science, (2) management, (3) society and (4) the individual.[1]

First Stage – Big Science-based Computerization

The first stage in the development of computerization took place in the period between roughly 1945 and 1970, which we will call the stage of big science. This refers to the period in which the computer began to be used extensively in national-scale projects, such as national defence and space exploration. In this case, the state (a subject) carried out computerization. Placing national prestige at stake, countries undertook the development of large-scale systems for national defence and for landing men on the moon. The United States took the lead, and used the computer extensively for SAGE (Semi-automatic Ground Environment System), protection against

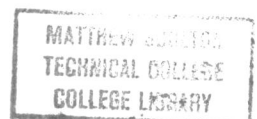

MATTHEW BOGGS
TECHNICAL COLLEGE
COLLEGE LIBRARY

Table 2.1 The developmental stages of computerization

Stage of development	First stage 1945–1970	Second stage 1955–1980	Third stage 1970–1990	Fourth stage 1975–2000
Bases of computer usage	Big science	Management	Society	Individual
Goal	National defence, space exploration	Gross national product (GNP)	Gross national welfare (GNW)	Gross national satisfaction (GNS)
Values	National prestige	Economic growth	Social welfare	Self-actualization
Subject	Nation	Organizations	General public	Individual
Object of computer use	Nature	Organization	Society	Human beings
Scientific base	Natural sciences	Management sciences	Social sciences	Behavioural sciences
Information object	Attaining scientific goals	Pursuing business efficiency	Solving social problems	Intellectual creation

Source: Japan Computer Usage Development Institute, 'The plan for information society', 1971

missile attack from the Soviet Union, in national defence, and for the Apollo programme in space exploration.

In the Apollo space programme the computer was put to use extensively in systems technology, in calculating trajectories and in remote control of spacecraft between the earth and the moon.

Second Stage – Management-based Computerization

The base of computerization in the second stage has moved from big science to management in both government and business. This stage extends from around 1955 to about 1980.

Unlike stage (1), in the second stage the expansion of GNP will be important. As computerization is applied to management and administration in both government and private enterprise, the computer will be used to improve the efficiency of operations of such bodies.

The development and use of management information systems (MIS), in which management and information sciences are linked, will advance.

In this second stage also, the United States has led the way, because the large-scale national defence and space information systems developed in the first stage have now been introduced into management and administration in big business and government. Big business in particular has made extensive use of the technology and systems developed by the government for national defence and space exploration. The SAGE system, for example, developed into the SABER system for passenger reservations in air travel, and the global inventory control system, developed for world military bases of the United States, has also been used by multinational corporations for inventory management on an international scale. Further, the methods of operations research developed during World War II have been used in management optimization.

With the development of methods such as systems analysis and PPBS, the computer has also come to be used by national and state governments as a powerful weapon in the making of complex policy.

Third Stage – Society-based Computerization

Computerization is now advancing to the third stage, society-based computerization, in which the computer will be used for the benefit of society as a whole. Society-based computerization has been advancing since 1970, and will probably go on through the 1990s, with many uses of the computer being applied to a *wide range of social needs*.

As this happens, GNW (gross national welfare) will become the goal instead of increases in the GNP (gross national product). The use of the computer will be for resolving problems in all areas of society, involving citizens as a whole. Thus *the general public will have a major role in the application of computerization at the level of the ordinary person*.

Of course, national and local governments will be responsible for determining policy in computer applications, but decisions regarding subjects will actually be made by the people, who will be directly affected by computerization at society level. At this stage, the social and interdisciplinary sciences, in combination with information networks, will be used extensively to solve complex social problems.

Take medicine, for example. New medical systems, such as medical care systems for remote places, and regional health management systems, can be brought under the computer system.

The educational system will be different from the present standard school education. *A knowledge network* will become the core of a new type of education, which places the emphasis on individual abilities. New social information systems will make use of the computer networks in a variety of social fields, covering matters such as pollution, traffic and problems of distribution. This third stage, therefore, has become the stage of computerization for the service of society at large, and in many countries this stage was entered during the 1970s.

Fourth Stage – Individual-based Computerization

Computerization is now entering its fourth stage of individual-based computerization, made possible by the invention of integrated circuits. It will probably enter this stage fully between the years 1975 to 2000. Computerization thus advances from the level of society at large to the individual level. Each person will be able to use computer information obtained from *man-machine* systems (systems in which dialogue between man and the computer is carried on by a conversational mode), to resolve problems and to pursue the new possibilities of the future.

At this stage, the information society will have reached the equivalent level of the most advanced stage of industrial society, the high mass consumption stage in which people in general have durable goods such as televisions and cars. The ready availability of information and knowledge will cause creativity to flourish among the people, the highest level of computerization, which I call *the high mass knowledge creation society*. At this stage there will be a personal computer in each household, used to solve day-to-day problems and determine the direction of one's future life. Such computerization will not be carried out by large organizations, but by individuals. *Each person will be the subject who carries out computerization*, and as this is done, great advances will be seen in the behavioural sciences.

I must add, in conclusion, that these four stages cannot be a series of mere successive developments, but each stage will continue developing even while the succeeding stage is coming into being.

Computerization from the Standpoint of Informaion Space

Having looked at the development of computerization according to each level of application, let us look at computerization from the viewpoint of information space and see how far it has already come. By 'information space' I infer the concept of the use of computer information spatially, referring to the range of a computer information network.

The expansion of this information space will go through three stages. The first will be that of limited space; the second, regional–national space, and the third, global space. The current stage of development in computerization can be comprehended by combining these three stages of development with the previous four stages of development in the areas of application. Table 2.2 sets out the developmental process for computerization in these terms.

First Stage – Computerization in Limited Space

The first stage, computerization in limited space, refers to restricted use of the computer by a business enterprise, a government, or a household, with the core technology advanced no further than the computer. Examples of this in big science are the calculation of missile trajectories by the military and the numerical calculations of atomic physics in universities and research centres. At management level, there are many examples of statistical tasks involved in compilation of a national census by the government, production control in factories, and the mass of routine calculations necessary to organizations. At society level, there are information retrieval services in libraries and CAI education in schools. And at an individual level, we can point to the existence of electronic calculators (the IC chip is a key component of the computer), home computers and TV games.

Second Stage – Computerization in Regional–National Space

The second stage of spatial development in computerization is in regional–national space. This refers to the range of information networks by which enterprises, government organizations, local governments and individuals carry on mutual exchanges and the

Table 2.2 Stages of computerization from the standpoint of information space

	Big science	Management	Society	Individual
Limited space (computer)	Trajectory calculations	Management information systems	Library information retrieval systems	Electronic calculators
		Numerical control systems	CAI education systems	Home computers
Regional-national space (computer + communications circuits)	SAGE	Ticket reservation systems	Coordinated traffic control systems	Push-button telephone service
		On-line banking systems	Regional pollution prevention systems	CATV systems
		Commercial TSS (time sharing system)	Regional medical care systems	Videodata systems
				Information utilities
Global space (computer + communications circuits + communications satellites)	Apollo space programme ERTS	Multinational management information systems	PEACESAT	Global information utilities
		World food information systems	Global medical care systems	

shared use of computer information. For this stage, the computer must be combined with communications circuits.

Examples include, at the level of big science, the early introduction by the military of national defence systems such as SAGE. The use of the computer by the military shows how suitable the computer has always been for use in networks. At the managerial level, there are the examples of the SAGE system being taken over by private enterprise in the passenger reservation systems of airlines and the on-line banking systems. The TSS (time sharing services) for business can also be included in this stage. And in Japan, in September 1970, society finally reached this level with the opening of the communications circuits to businesses. Examples at the society level are the coordinated traffic control systems, pollution monitoring, prevention and elimination systems, and regional emergency medical care systems. The formation of regional–national information space at the level of society is just coming into being in many countries.

Finally, at the individual level, there are the examples of the push-phone, CATV, and the *wired city*. Experimental projects of community communication visual information systems began in Japan in the Tama New Town and Higashi Ikoma projects of 1976.

There are also *videodata systems* by which subscribers are able to dial up a computer by telephone and have information displayed on their own TV screens. This was originally in the United Kingdom in 1978, and many other countries, including the United States, Canada and Japan, have since adopted these in concert.

These movements suggest that the era of information utility is really at hand.

Third Stage – Computerization in Global Space

The third stage refers to computerization in global space. Communication satellites are added to the information technology of computer and communications circuits. At the level of big science, this stage has already been entered. The series of back-up systems for the Apollo programme and ERTS (Earth Resources Technology Satellite) are typical examples. ERTS is proving to be particularly effective in the search for underground resources. At the managerial level, a global information service using communication satellites has already been put into operation by General Electric, and many American multinational corporations have begun to operate their own global management information systems. At the level of society, computerization has not yet escaped from the experimental stage, and has scarcely moved beyond PEACESAT (Pacific Education and Communications Experiment by Satellite), an experiment in implementing an educational information system in the Pacific Basin.

On an individual level, computerization appears still to be no more than a phantasmal picture of the future. But as we enter the twenty-first century *global information utilities* will become realities, and people in any part of the world will be able to utilize these GIUs as freely as we now use international telephones.

It will be possible for people all over the world to obtain services ranging from self-education systems, library information systems, to enjoy competitive games, and to *participate in a global voting* system to deal with such issues as atomic power generation.

A Comparison of the Tempo of the Motive Power Revolution and the Information Revolution

Let us seek a clearer idea of when the information society will come into being by using various indices to make a comparison between the tempo of the motive power revolution of industrial society and the information revolution of the information society. The comparison has shown that the latter revolution has occurred three to six times faster than the motive power revolution; it can be predicted that the high mass knowledge creation society, the most advanced stage of the information age, will probably be actualized somewhere on earth by the middle of the twenty-first century (see table 2.3).

Let us look at the dates of technological innovations that have played decisive roles in the power revolution. Thomas Newcomen invented the first steam engine in 1708, and James Watt improved the Newcomen engine and completed the first operable steam engine in 1775. The first railroad was laid between Liverpool and Manchester in 1829, and the production of the Model-T Ford, the first car on the mass market, began in 1909. The jet plane emerged in 1937. There was a period of 229 years in technological development between the invention of the Newcomen engine and the emergence of the jet plane.

Note the contrast in the computer revolution. ENIAC, the first vacuum tube computer (first generation computer), was developed by Eckert and Mauchly in 1946. The second generation computer, using transistors, was developed in 1956; the third generation computer, which utilized integrated circuits, appeared in 1965. Microprocessors, integrated circuits on one chip, appeared in 1973, and now the development of VLSI (very large scale integrated circuits) is under way. Success seems assured, and the fourth generation of computers came into practical use early in the 1980s.

Technological developments in the motive power revolution from the Newcomen to the jet engine took approximately 230 years, but in the information revolution, the period from the first generation to the fourth generation computer will probably not be more than some 36

Table 2.3 A comparison of the tempo of the motive power revolution and the information revolution

	A. Motive power revolution			B. Information revolution			Ratio A/B
Advancement of technology	Newcomen engine	1708 ⎫		First generation computer	1946 ⎫		
	Steam engine	1775 ⎬ 229 years		Second generation computer	1956 ⎬ 36 years		
	Railroad	1829		Third generation computer	1965		6.4:1
	Automobile	1909		Micro-processor	1973		
	Jet plane	1937 ⎭		Fourth generation computer	1982 ⎭		
Diffusion of core machinery and systems	1,500 Steam engines 1708 ⎱ 92 years 1800 ⎰			30,000 Computers 1946 ⎱ 20 years 1966 ⎰			4.6:1
	1,000 Power looms 1784 ⎱ 49 years 1833 ⎰			Automatic data processing 1946 ⎱ 9 years 1955 ⎰			5.4:1
Industrial development	Construction of American transcontinental railroad 1828 ⎱ 41 years 1869 ⎰			Formation of an American nationwide information network 1965 ⎱ 7 years 1972 ⎰			6.0:1
	Establishment of manufacturing industries 1708 ⎱ 201 years 1909 ⎰			Establishment of information industries 1946 ⎱ 44 years 1990 ⎰			4.6:1
Societal development	High mass consumption society 1708 ⎱ 222 years 1930 ⎰			High mass knowledge creation society 1946 ⎱ 64 years (?)2010 ⎰			3.5:1

years. The information revolution will have occurred about 6.4 times faster than the power revolution.

Looking at the diffusion of the core machinery and systems in each case, we find that the motive power revolution took 57 years for the development of the Newcomen engine to reach 1,000 units, and a further 35 years for James Watt's engine to spread throughout modern industry, such as steel, coal mining and spinning. Together, this adds up to 92 years. In 20 years from the development of the computer, more than 30,000 machines were put into operation throughout the world. The diffusion of the computer has taken place about 4.6 times faster than the diffusion of steam engine power.

Take another example. It took 49 years for the first 1,000 power-equipped spinning machines to be sold, but it took only nine years for data processing by computers to be introduced into business. Automatic data processing has moved 5.4 times faster than the diffusion of power-equipped spinning machines.

When we look at industrial development, we see that it took 41 years for railroads to achieve the construction of the American transcontinental railroad, but only seven years were needed for an information network to cover the same American continent, six times faster than the railroads. If manufacturing industries became 'leading industries' with the establishment of the automobile industry, then the development from the Newcomen engine took 201 years. The 'leading industry' to be developed from the computer will be the information industry. While the information industry is still in the process of formation, we can expect that its position as an industry will be established by about 1990, which means a development over 44 years, which is 4.6 times faster than the development of the manufacturing industries.

Finally, let us look at societal development. We have referred to the most advanced stage of industrial society, brought into being by the motive power revolution, as the high mass consumption society. If this is regarded as coinciding with the spread of motorization in a high mass consumption society, then this stage was reached in the United States in 1930, 222 years after the introduction of the Newcomen engine. It will take quite some time for the high mass knowledge creation society to emerge as the most advanced stage of the information society. If this is regarded as coinciding with the joint utilization of information utilities by the people, this stage will be reached by the end of the first decade of the twenty-first century, about 3.5 times as fast as the development of industrial society.

It should be clear now that whichever of these four indices one uses, the speed of the information revolution will be between three and six times the rate of development of the motive power revolution.

Notes

1 Y. Masuda, 'The Conceptual Framework of Information Economics, *IEEE Transaction on Communications*, New York, IEEE Communications Society, October 1975.

PART II

Framework of the Information
Society

3

The Information Epoch: Quiet Societal Transformation

Why Computer-communications Technology will bring about an Information Epoch

Mankind is in the process of the fourth information epoch[1] centring on computer technology operating in conjunction with communications technology. This information epoch centring on computer technology will have a far more decisive impact on human society than the 'power' revolution that began with the invention of the steam engine. The basic reason is that the fundamental function of the computer is to substitute for and amplify human mental work, whereas the steam engine had the basic function of substituting and amplifying physical labour.

The world's first computer, ENIAC, was invented by J. P. Eckert and J. W. Mauchly in 1946, 171 years after James Watt's steam engine of 1775. The invention of ENIAC was motivated by military needs, the need for high-speed calculation of flight characteristics of projectiles for military purposes. This machine was essentially different from the tools or machines that had so far been invented, in that it had a mechanical calculating brain.

The importance of computer technology is in the fact that *for the first time a machine was made to create and supply information*. The computer was an epochal machine of logic, equipped with the three information processing functions of memory, computation and control, which greatly increased human ability to originate information.

The Complete Objectification of Information

The computer has three tremendously superior characteristics as a man-made intelligence machine. The first is *the complete objectification of information*. This means (1) production of information independent of human beings, (2) the originality of information thus produced, and (3) its storage in preservable forms. It can be said that this objectification of information is an index to progress in the structure of information. Mankind carried out the first information revolution as a revolution in language, but at this stage, information could not be objectified, which means *the separation of information from its subject*. Information was simply transmitted from A to B and had not become independent of man. Objectification of information began with the advance of production technology that brought it to the stage of written information. In the forms of alphabet and ideograms man first inscribed information on stone, which was thus transmitted to a third party in a completely independent, objectified form. This can be called *primary objectification*.

When the information revolution reached the stage of the printing revolution, objectified written information was disseminated in multiform reproduction by the printing press, and became typographically recorded information. This meant that information was objectified at a second level, with the shift from written to typographical information, *the secondary objectification*. As the information revolution progressed, the tendency towards separation and objectification of information increased further. The computer revolution has not simply advanced this objectification of information another step; it has carried out a critical qualitative leap by completely separating the production of information from the subjects, so that the production of information moves from man to machine. I call this objectification *the tertiary objectification* of information. Figure 3.1 sets out this process. As the figure shows, primary objectification occurred at the stage of written information, the secondary objectification took place at the stage of printed information, and the tertiary objectification was initiated with the electronic processing of information. At this stage the production of information by a machine begins, and the complete objectification of information is achieved.

The process of objectification of information may be likened to mankind's progress in knowledge. The invention of written characters as the first form of objectification of information made it possible for

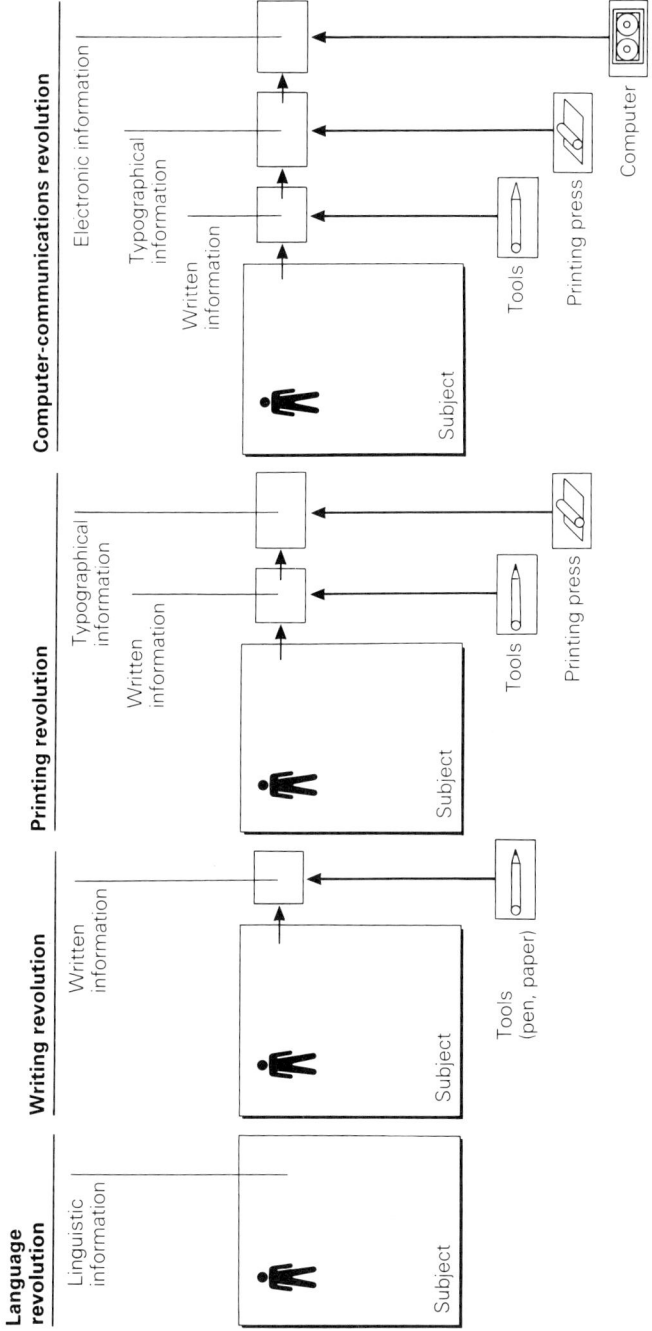

Figure 3.1 Objectification Process of Information.

the first time in human history to preserve information in the form of objective existence. This led to the systematic development of technology and the formation of a composite body of knowledge, and the creation of poetry and literature. Further, the development of printing as the second-stage objectification of information led to the copying of information by machines, making it possible greatly to speed up the production of information, bringing in its wake the mass production and mass distribution of information and knowledge, the popularization of literature and knowledge, and the widespread dissemination of information by the press and other mass media.

Furthermore, the invention of the computer as the third objectification of information has made it possible to produce original information by a machine. The computer also has the functions of memory, calculation and control, operating as a mechanical brain. The development of the computer can thus replace the production of information by man, by automating the production of information and substantially amplifying man's intellectual capacity through man-machine type utilization of computers.

Production of Sophisticated Cognitive Information

The second great characteristic is *the production of sophisticated cognitive information.*

The computer does not merely produce information; it produces sophisticated *cognitive information.* Here 'cognitive information' refers to information that is *a projection of the future*; it is *logical*, and it is *action-selective.* The projection means that the cognitive information is used for detecting and forecasting. 'Logical' signifies *the existence of a goal and ends relationship; cause and effect within cognitive information,* and 'action-selective' means that the information is used for *the selection of actions and means most appropriate for achieving a goal.* The fact that the computer can produce this cognitive information mechanically and in large quantities is a great contribution to the amplification of mental labour.

The primates, including man, make full use of two kinds of information to sustain and enrich the content of their living. One is cognitive information and the other *affective information.* 'Affective information' refers to information that is based on *sensitivity and production of emotion.* It embraces all the information that conveys sensory feelings, such as 'comfort', 'pain' and the emotional feelings of 'happy' and 'sad'. These two kinds of information operate like two wheels of a car, and both are essential for human living. But it is the knowledge-oriented information that plays the decisive role in the progress and development of human society. 'Emotional information', to put it precisely, is from the beginning an expression of emotional life that is *simply for an organism's own use satisfaction*, not the sort of information that brings about an active change in the situational relationship between a subject and the object around it.

'Cognitive information', on the other hand, *makes possible purposeful action selection in response to changes in situational relations*. Cognitive information, in other words, makes possible the active interaction with the external environment, resulting in change, from which progress takes place in the life of mankind.

The Information Cycle

In protozoans there is a prototype of cognitive information, which seems to indicate that cognitive information came into being with the very origin of life. The fundamental characteristic of an organism is that it takes in food to sustain its existence, and carries on reproduction in order to continue the species. Thus an organism seems predestined to be forever acting on the external environment, looking for food, and protecting itself from external enemies. Here cognitive information plays an indispensable role.

A *situational relation* exists between the organism and the environment surrounding it. The 'situational relation' is *the relation between a subject and an object that comes into being in a particular situation*, and to which there are three conditions: (1) there must be a subject and an object (in this case the environment surrounding the subject), (2) the subject must receive impulses from the object, and (3) the subject takes action in response to these impulses. More noteworthy, however, is the pattern of action that is visible in the living protozoan. A protozoan uses organelles to ingest food and reject alien matter in response to signals received from outside. The four continuous steps that can be seen in the simplest actions of protozoans, *subject–object–signal–action*, constitute the fundamental structure of cognitive information, which we can call 'the information cycle'.

The *information cycle* is shown in figure 3.2. The subject receives a signal from the object, identifies the signal and evaluates it according

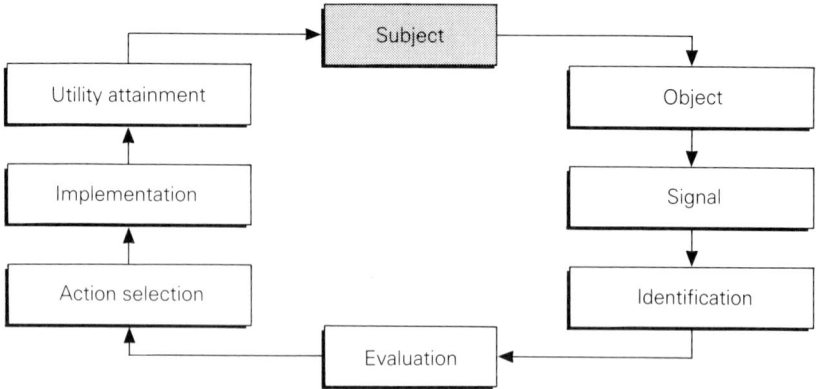

Figure 3.2 The Information Cycle.

to an acquired standard of judgement, selects a course of action, and finally achieves some use-value by implementing the action.

The Function of Information

In order to understand the information cycle more clearly, we can take the example of the amoeba, and examine it in detail. In this information cycle the subject (the amoeba) first receives signals from the object (external environment) by means of its organelles. There are four types of signals provided by nature: physical and chemical, temperature, light, acid and alkali. The amoeba does not simply receive these signals; it distinguishes between them qualitatively and quantitatively.

The amoeba responds to change in the frequency of these signals when it selects a course of movement and transfers this into action. It moves, for example, in response to a change in temperature indicating warmth or cold. It distinguishes between differences of temperature as subtle as one-twentieth of a degree per cubic centimetre of water surrounding it, and accordingly changes the direction of its movement. Obviously the amoeba has some sort of standard by which it evaluates the quantitative changes of these signals. This becomes even more apparent when one looks at the amoeba's intake of liquid food; it takes in liquids as well as solids. The liquid it selects to ingest is according to the type of liquid. It also takes in salts and proteins and expels carbohydrates and nucleic acids. The concentration of salts which can most readily be absorbed is one-tenth of a mole.[2]

The Utility of Cognitive Information

Let us look at the process by which cognitive information is utilized. We have seen that (1) the amoeba selects some actions (action selection) to take in food when a signal indicates that food is there, and that (2) successful implementation of this action secures life support for the amoeba. The value of the signal the amoeba receives is that it is useful in selecting an action that will maintain the life of the individual amoeba and the continued existence of the amoeba as a species. Another way of putting it is that *the utility of cognitive information* is attained when the action the subject has taken *brings about a change in the situational relationship between the subject and object*, enabling the subject to attain its goal. When a signal indicates to the amoeba the presence of food, the amoeba achieves the goal of selective action when it ingests the food. The situational relationship between the amoeba and the food changes, and the use of cognitive information is a completed act.

The process, then, to realize the utility of cognitive information is structurally *the goal-oriented feedforward* of a subject itself, and of the environment by the subject.

We are accustomed to the word 'feedback', which refers to *returning a deviation back within the control boundaries*. In 'feedforward', *a goal exists, but the boundaries of control are not fixed, and must be adjusted dynamically to a changing situational relationship*.

Usually, a subject must go through several information cycles to achieve the situational relationship and the use-value of cognitive information that is its final goal. As each information cycle is completed, the subject achieves an intermediate level of utility, and a new situational relationship is established. The next signal the subject receives, thereby starting a new information cycle, stems from this new relationship. This process is repeated over and over until the subject achieves its final goal. The utility of cognitive information, therefore, is a result of a goal-oriented feedforward of a subject itself and its environment through the action of a subject to achieve a goal. Here goal-oriented feedforward is control in the direction of a goal. In response to changes in situational relations the subject controls its actions, and through its actions controls the external environment.

We can now give a more complete definition of 'cognitive information': it is *an informed situational relation between a subject and an object that makes possible the action selection by which the subject itself can achieve some sort of use-value.*

This definition is sufficiently tenable as a general definition of information, if information is understood to be something counterposed to knowledge and technology.

'Knowledge' is no more than *cognitive information that has been generalized and abstracted from an understanding of the cause-and-effect relations of a particular phenomenon occurring in the external environment.* 'Technology' is *cognitive information that is useful in effectively carrying out production-oriented labour requiring a certain degree of prescribed expertise.* Fire-lighting technology in primitive society refers to production-oriented labour based on experience, such as 'rubbing wood together will light a fire'. Many repetitions of isolated instances of lighting a fire were necessary. In the beginning, empirical fire-lighting information was acquired by mere chance, and as information was accumulated from experience, it was passed on to others. Finally, standardized technological cognitive information became firmly established in human society, in what we know as technology.

Formation of Structurally Organic Information Networks

This is a third great characteristic of the information networks made possible by the combination of computers with communications technology. An *information network* is seen in the transmission of information between a large number of people within an extensive area made possible by the telephone and telegraph networks. This network, combined with a computer, has been developed into a network system that closely resembles information mechanisms as a living body, an organism.

A system of information in such a living body, the man-machine combination, can be broadly divided into two types of systems, *environmental* and *organismic*. These two systems of information are in a sense complementary sub-systems of information for the maintenance and development of the existing organism. Each has a different function.

The system of 'environmental information' is concerned with *the relation between an organism and the external world* in order to maintain its existence. The fundamental functions of the system of environmental information include such as to enable the organism to catch food and protect itself from enemies in order to survive. The system of 'organismic information' is concerned with carrying on *the essential functions within the living body* of the organism itself. The main function of the system of organismic information is to carry out physiological functions to maintain life within the body of the organism. The important point to note is that a primitive prototype of *the system of environmental information emerges from the highly developed system of organismic information.* Systems of organismic information are astonishingly intricate, yet are already wonderfully complete in protozoans. Be that as it may, what is important to stress here from the standpoint of the systems of information in organisms is that the existence of the intricate system of organismic information first made it possible for the system of environmental information to come into being.

Comparing the systems of organismic information in protozoans with the systems of environmental information in mankind that have come into being with computer technology reveals a number of similarities. The first is that both have *a memory function.* The protozoan has DNA that serves this function, and the computer has magnetic tapes and drums. In the protozoan the genetic code is inherited. There was no such thing as memory information until cognitive brains evolved in mammals. In this respect the computer is far ahead of the protozoan.

The second similarity is that both have *a programmatic function.* Common to both the protozoan and the computer is that they are provided with a program, and cannot produce it themselves. What is different is that while the protozoan's program is hereditarily fixed in the DNA as a genetic code, man creates a program for the computer for each time of usage.

The third similarity is that both have *a copy function.* In the protozoan a genetic code is copied from DNA on to RNA. In the computer, programs and data can be copied from the magnetic memory on to the CPU (central processing unit) of the computer.

The fourth similarity is that both have *a coding function.* In protozoans, the genetic code is retained in the form of unique genetic signals that indicate each of the characteristics of the four nucleotide bases (adenine, guanine, thymine and cytosine), the component molecules of proteins and enzymes. A sequence of three nucleotide bases on an m-RNA strand (called a 'codon') indicates the amino acids that make up proteins and enzymes, and these are translated from m-RNA by t-RNA. In the computer, numbers and letters are coded into combinations of o 1 bits, the information is processed in this state, and then these bit codes are translated back again into numbers and letters as output.

The fifth similarity is that both have *a control function* – discrimination, evaluation and command. In the protozoan, RNA (a compound enzyme) has this function. The t-RNA discriminates between the amino acids in the food taken in, according to the combination of genetic signals that the t-RNA of the protozoan has copied, and brings each of these amino acids together according to a program. In this process the functions of discrimination, judgement and evaluation are clear. Also, t-RNA is equipped with the coded signals 'start' and 'stop' in the program.

In the same way, the computer discriminates between and judges the information it has recognized. Based upon this, it can give instructions and control commands. It also signals 'start' and 'stop'.

Looked at in this way, the system of environmental information that constitutes the information structure of the computer has begun strongly to resemble the system of organismic information. When the computer is joined with communications technology, the system of environmental information approaches the systems of organismic information of more sophisticated organisms, seen in three ways.

The first is the development of *the system of information transmission and control*. Similar to animals, which have a control system of motor and autonomic nerves extending throughout the body, the computer, with its communications lines, constitutes a cognitive information network, and makes possible networks of simultaneous concentration and dispersal of cognitive information. Data from many places cannot only be concentrated and processed in one place, they can also be simultaneously dispersed to many places.

Moreover, response networks came into being. Simultaneous concentration and dispersal of information processing began to be carried out continuously to provide instantaneous responses, and complex feedforward networks became the most sophisticated function of cognitive information networks. It became possible not only to present goal-directed feedforward continuously, but also to process and transmit many different types of feedforward information in a complex way.

The second similarlity is *the development of diverse information organs*. In animals, information organs such as eyes, ears, mouth and nose are developed; in computer information networks, many different kinds of input and output apparatuses (line printers, cathode ray tubes, facsimile, intelligent terminals) are the terminal equipment.

The third similarity is *the development of advanced information processing technology*. In a human being, even a seemingly simple action like walking first becomes possible when there is information exchanged within a network of bodily organs, including, of course the eyes, muscles of the legs, and the equilibrium organs of the middle ear. For computer networks there are also sophisticated means of information processing. On-line real-time systems provide instantaneous response over long distances. Through time sharing systems many people can use the computer at the same time. Teleconferencing uses television and the computer network to conduct conferences, without people having to assemble in one place. The remote sensing technology that resulted from joining the computer with communication satellites has permitted the development of a system of searching for weather conditions and resources allocation on a worldwide scale.

Possibility of Highly Organismic Society

The similarity of the system of environmental information of computer-communications technology to the system of organismic information in living organisms suggests something important for the outlook of the future information society. The hypothesis can be formulated that the future information society will be a 'highly organismic society' resembling an organism. What I am referring to will probably be *a multi-centred complex society* in which many systems are linked and integrated by information networks. Moreover, this society will have the dynamism to respond more quickly and more appropriately than contemporary society to changes in the external environment, and then the information society of the future will appear before us as a society with highly organic information space linked by a network of cognitive information with complex feed-forward loops.

The Societal Impact of the Information Epoch

The information epoch to be brought about by computer-communications technology will not simply have a big socio-economic impact upon contemporary industrial society; it will demonstrate a force of societal change powerful enough to bring about a transformation into a completely new type of human society, which is the information society.

Generally speaking, innovational technology changes social and economic systems through the following three stages:

1 In which technology does work previously done by man.
2 In which technology makes possible work that man has never been able to do before.
3 In which the existing social and economic structures are transformed into new social and economic systems.

The three stages of technological innovation, as they apply to the revolution in computer and communications technology, may be defined as *replacement and amplification of the mental labour of man and the transformation of human society*. This may be defined as follows:

1 The first stage is *automation*, in which man's mental labour is increasingly accomplished through the·-appliction of computer-communications technology.
2 The second stage is that of *knowledge creation*, which entails the amplification of man's mental labour.
3 The third stage is that of *system innovation*, a set of political, social and economic transformations resulting from the impact of the first two developmental stages.

A brief outline of the societal impact of the computer-communications revolution is shown in figure 3.3.[3]

Automation: the Replacement of Man's Mental Labour

Automation has traditionally been understood to involve the assumption of various kinds of mental activities (recognition, understanding, computation, memory, judgement, control, etc.) by computers or computer-driven servo-mechanisms. However, the expanding applications of computer-communications technology are changing this traditional concept to the extent that the future prospect for computer-oriented automation has greater potentialities than we commonly accept today.

First computer-communications technology will bring about the complete automation of production. Industrial production is simply

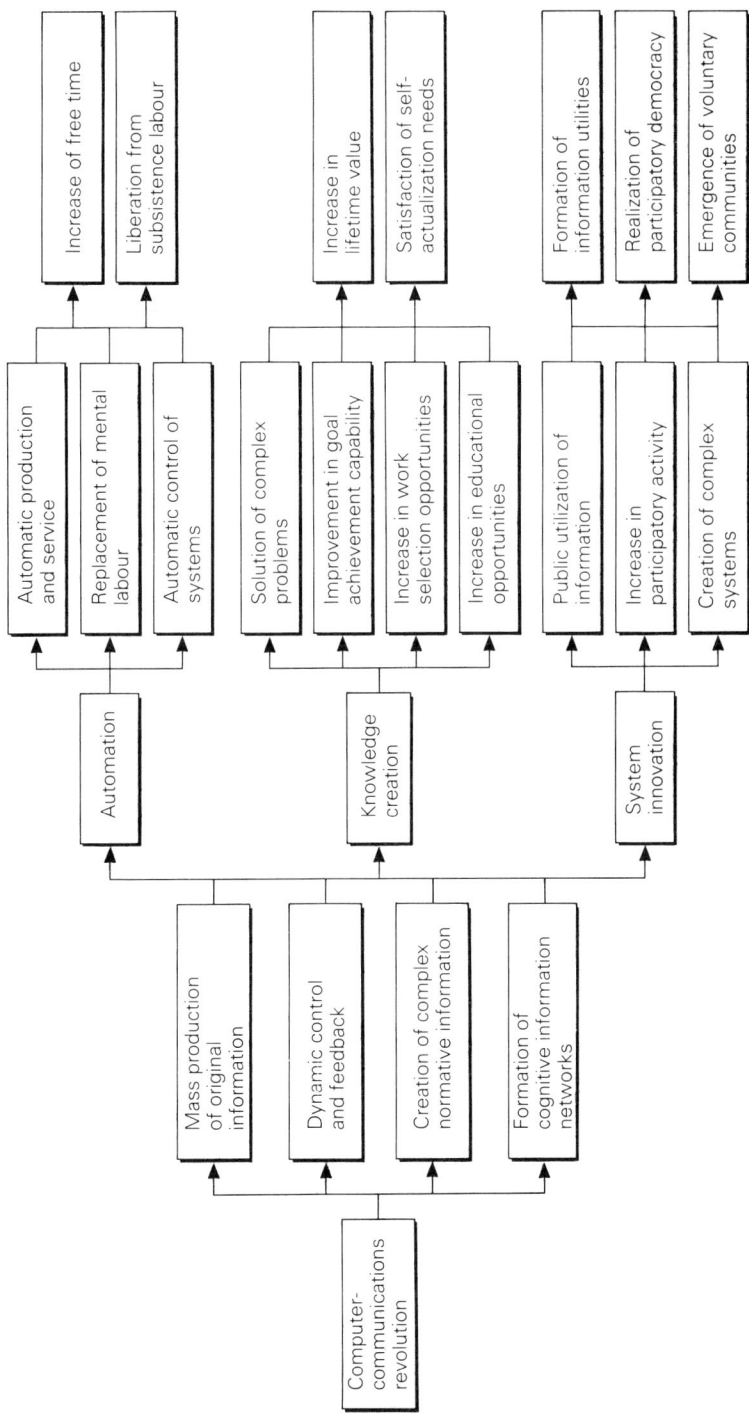

Figure 3.3 The Computer-communications Revolution and its Societal Impact.

the process of applying scientific laws to change raw materials into useful goods, and the computer functions to feedback the production process quickly in response to such changes that occur during the production process. Computer automation began with evaluation and control, and went on to group control systems and the complete automation of a whole production process. In the near future, complete automation of entire plants will come into being, and during the next twenty or thirty years there will probably emerge, in fields related to energy and materials (electric power generation, oil refining, iron, cement, etc.), factories that require no manual labour at all.

Second, computer-communications technology will bring about *automation of knowledge-oriented services and operations*. Whenever man's knowledge-oriented activity is carried out in a fixed logical order, a computer can be programmed to perform in the same way. In knowledge-oriented services, many kinds of automated service machines, such as vending machines and cash dispensers, are already replacing the service labour of man. Even in medical treatment, automatic diagnosis devices, as used for examining patients suffering from cardiovascular ailments, have already been developed. In knowledge-oriented operations, clerical duties related to billing and accounting have already been completely replaced by the computer.

Third is *systems automation*. This is the automation that creates unified systems that combine many sub-automation functions organically in place of separate and independent feedback and control. The group numerical control systems of industrial production are typical examples of systems automation. Systems automation makes possible large scale integrated traffic control systems to optimize the flow of traffic through hundreds of intersections. Systems like the lunar landing system using satellites to put man on the moon, of course, are semi-automatic man-machine systems that utilize computer-communications technology more than any other.

In discussing the social impact of automation, however, it is necessary to deal with the existence of both bright and dark potentials.

A first principal social impact will be *the increasing emancipation of man from labour for subsistence*: this will have an immeasurable social and psychological effect on the future of mankind, a social impact which may be said to belong to the bright side of automation. Ever since his appearance on earth man has devoted the great bulk of his time to work merely to assure his physical survival. In this, man has been bound in space; that is, to the physical places of production. In the agricultural age, man was bound to the earth, and in the industrial age, workers were virtually confined to factories and office workers to their offices. However, since a large portion of office and production work will be replaced by computers, there will be less need for men and women to serve as the medium of information storage and communication, which is, after all, the principal reason for human involvement in much organized work. Thus, man will not only be emancipated from the necessity to labour for subsistence, but he will also be freed from the bonds which have tied him to the places of production. Moreover, man will increasingly have more time to expend on his personal satisfactions.

In the light of these changing circumstances, it would be unreasonable to assume that the traditional concept of leisure will continue unchanged in an information society. It would be more reasonable to suppose that a new concept, suitable to the information epoch, will take the place of labour for subsistence. In this context, I prefer to refer to 'leisure' in the information age as 'free time'. Free time is not merely the opposite of work time; rather it is all time which may be disposed of freely by each individual. *Free time* may therefore be broken down into three parts.

The first is leisure in the conventional meaning of the word, the content of which is *rest* and *play*. But rest and play will not be able to fill our ample free time in the future.

The second part of free time would involve learning in the broadest meaning of the word. In other words, more and more people will spend their free time by studying systems sciences and computers in order to adapt themselves to the information epoch, or by taking lessons in cultural accomplishments, hobbies, arts and crafts. (In the United States, adult schooling is already the fastest growing form of education.)

The third segment of free time will involve preparation for a better life in society, or for collecting and analysing information for social activity and for working on plans for the future, and making projections.

If the freeing of man from subsistence labour is a positive social result of the information revolution, a second social result, *unemployment*, will represent a negative side of automation. There is ample reason to fear that the unemployment of old and middle-aged persons and the obsolescence of old techniques as a result of automation will pose a serious social problem. This is an unavoidable choice between either an increase in free time or mass unemployment, a choice that the future information society will have to face. I feel that, just as in the past, developments in industrial productive power have ultimately brought about an increase in consumption and income, not an increase in unemployment, the fruit of automation resulting from the development of information productive power in the near future will probably be an increase in free time, rather than unemployment.

A third social impact of automation is that of *social restraint*. This not only represents the darkest side of automation, but is also perhaps the most critical issue of applied computer-communications technology. While liberating us from labour for subsistence and providing us with ample free time, automation will bring the possibility of *invisible social restraint*, so called because it would not entail surveillance by secret police.

One such source of alienation would be related to man's restraint by functions or systems. As we have noted, automation will liberate us from the place of production and restrictions in time. Businessmen, for example, will be freed from the need to use packed commuter trains; their independent initiative will fill a higher role in their work. On the other hand, they will be more strictly subjected to management goals based on merit and performance. Further, management information systems (MIS) will relate each person's activities closely through various functions of management, and on-line, real-time control systems will establish very strict time schedules for these functional relationships. When such automatic management control systems become commonplace, functional and systematic restraints will replace the restrictions of place and time.

Another type of alienation will centre upon the potential for *the invasion of privacy*. In an advanced information society, there will be thousands of data banks of various kinds which will contain massive information data on individuals and enterprises. For instance, a government social data bank will accumulate information according to a unified numbering system on all individuals with respect to their numerous activities from the cradle to the grave, while financial institutions will have detailed information on the balance sheets of every family. The use of such private data by government or by any other group of persons for a particular purpose would constitute a serious cause for the further alienation of man. Of course, laws will have to be formulated to prevent such invasions of privacy, but such laws could prove inadequate for the purpose.

Further alienation may result from the use of computer-communications technology to create *a managed society*. The managed society would operate in such a way that ruling elites would guide the 'managed' (persons and things), using information networks as control mechanisms. Information created by computers is confined to quantifiable and logical information; thus, much information regarding human aspects of life cannot be computerized. It therefore follows that a computer-managed society may become inhuman, or alienated from humanity. A completely automated state would be an intellectual ice age, devoid of humanity, in which a handful of data manipulators would dominate as an intellectual elite.

While recognizing these dangers, however, I believe that mankind will be able to avoid such a managed society, and be able to proceed to the second stage of development: knowledge creation. My reasons for confidence are taken up later.

Knowledge Creation: the Amplification of Man's Intellectual Labour

If automation entails the supplanting of man's mental labour by computer-communications technology, then *knowledge creation* is a concrete example of the amplification of mental labour through such technology.

By 'knowledge creation' is meant *the creation of intellectual values*, but this is only a commonplace and general definition, and is illusory. We can grasp it by seeing its two aspects, viz., *problem solving* and *opportunity development*. In this context, 'problem solving' is devising *a method or means of eliminating risks that may stand in the way of accomplishing an aim*. Computer-communications technology can help us expand our problem solving capacity by reaching beyond the limitations of time and space, a capacity found in the very technological foundations of computers – large memory capacity, high-speed calculation, integrated control functions – in an on-line, real-time system.

One of the most advanced problem solving systems of this kind is the forecasting, evaluating and warning system, a system for quick discovery of problems in rapidly changing circumstances, forecasting a future trend, evaluating the degree of danger due to these problems, and issuing a warning when danger appears.

Problem solving systems of this type have some fundamental characteristics.

The first is that they are *pro-active*: a system for detecting a problem before it becomes serious, and by predicting future trends, projecting potential alternative solutions. The second is for discovering hitherto unknown problems. The third characteristic is that the problems to be solved by systems of this kind are very complex, protracted and not straightforward (e.g. environmental disruption by chemicals).

Among the large-scale forecasting, evaluating and warning functions the system is certain to provide in the future information society is ecological forecasting, evaluating and warning. Vast numbers of measuring instruments will be installed in all parts of the globe, from the Arctic to the Antarctic, to be used in conjunction with space observation statellites, all of which will be connected to regional centres. The regional centres will be connected to a United Nations centre, where a computer, many times larger in capacity than conventional computers, will operate for this global ecological system. The UN centre will supply constant information on weather conditions, air, sea and river pollution in different regions of the world, and on the basis of the information received, forecasts and warnings will be issued.

Such a vision is far from being unrealistic, and the UN is already working on such a system, the major problem of which is not technical. To implement it fully calls for readiness and a cooperative attitude of UN member nations, and a change in their philosophical thinking on the future of humanity, and in their sense of values.

The second aspect of knowledge creation is *opportunity development*. By 'opportunity development' is meant *research and development of possibilities of future time usage or creating new values in rapidly changing environmental conditions*. Opportunity development is encouraged by the existence of the *information utility*, which will come into being when information becomes a public commodity, similar to water and electricity, which one can obtain as needed. This is the important societal orientation of computer-communications technology. The first societal impact of the information utility is *an increase in the opportunity for education*.

With the development of the information utility, one will be able to acquire and use cognitive information at any place and any time; this means that education will be freed from the restrictions of income, time and place. The result will be that all human beings will have the educational opportunities they desire, with conditions that make it possible for them to develop the full potentialities of the future.

The second impact will be *the increase in opportunities for work.* Through the information utility, people will be able to obtain much more information more quickly than now, relating to the possibilities for new work. People will have many opportunities for choice when selecting future work or the direction of their social activity. A new industry, *the opportunity industry*, will develop in response to the need of individuals and groups for opportunities for development. The opportunity industry will aim to help individuals and groups develop and realize their future potentials.

The main sectors of the opportunity industry will be the education industry, the information industry, the mass communication industry and the consultation industry; and industries concerned with psychosomatic medicine and molecular biology, for example, will also have their part. There may even emerge something resembling religious activities, as we see religion in various forms again becoming a day-to-day factor in life in the twenty-first century.

System Innovation: Emergence of New Socio-economic Systems

The third aspect of the information epoch will be *system innovation.* This means that present socio-economic systems will be replaced by new socio-economic systems. System innovation wil be the most far-reaching effect of the information epoch.

When epoch-making technological innovation occurs, changes take place in the existing society and a new society emerges. The steam engine precipitated the industrial revolution, bringing about changes that led to a new economic and political system: the capitalist system and parliamentary democracy. The information epoch resulting from computer-communications technology will bring about a societal transformation just as great or even greater than the industrial revolution.

Let us look at some typical examples of major and basic transformations to be expected: a change in our values system from material to time value; from a system of free competition to a synergetic economic system; from parliamentary democracy to participatory democracy.

These matters are discussed in detail later; let us focus on the transformation in the educational system, as a most dramatic societal change.

The first change will be *to lift education out of the restrictions of formal schools*. The present closed educational environment will be replaced by an open educational environment, made up of *knowledge networks*. It will eradicate the educational gaps between town and countryside, and between industrial and non-industrialized countries.

The second change will be *the introduction of a personal type of education*, suited to the ability of each individual, replacing the traditional uniform system of collective education with a system determined by individual ability and choice. This will become possible through educational programmes suited to different levels of scholastic attainment with a wide range of educational opportunities. This means that the present educational system, graded according to age, will be supplanted by a system that allows people's abilities to move on to advanced courses, irrespective of age, and where even children of lesser ability will be able to improve their levels of learning by means of personal-type lessons and guidance.

Thirdly, *self-learning* will become the leading form of education. The formal educational system has been one of unilateral teaching of students by teachers. When a system of self-learning is introduced, teachers will act as advisers or counsellors. This will be possible because, as a result of the development and spread of CAI (computer-aided instruction) systems, students will be able to study by themselves, watching CRT displays and conversing directly with a computer and with other people by computer.

The fourth change will be to *knowledge-creative education*. Education in this industrial society aims at cramming the heads of students with bits of information and training them in techniques. This will be replaced with knowledge-creative education and training, because the information society will develop through information values into a high knowledge-creation society.

The fifth change will mean *lifetime education*. The present education system is centred on compulsory education to be completed when young. There are few higher and professional opportunities of learning available to the average person after that. In the information society, however, greater importance will be attached to the education of adults and even elderly people, because this will be necessary to enable adults and elderly people to adapt themselves to the changes of the information society, and to develop their abilities for society as a whole to accept the increasing proportion of elderly people in the population.

This radical change in the educational system will be of great significance to the development of human history – a historical transition from industrial society, in which the natural environment has been unilaterally transformed and material consumption expanded, to the information society which seeks coexistence with nature through mankind's own transformation and innovation of new socio-economic systems.

Notes

1 S. Kuznets, *Modern Economic Growth: Rate, Structure and Spread*, New Haven, Conn., Yale University Press, 1966.
2 J. Ota, *Amoeba*, Tokyo, NHK Books, 1960.
3 Y. Masuda, 'Computopia vs. Automated States: Unavoidable Alternatives for the Information Era', *The Next 25 Years: Crisis and Opportunity*, Washington, DC, World Future Society, 1975.

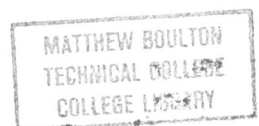

MATTHEW BOULTON
TECHNICAL COLLEGE
COLLEGE LIBRARY

4

Globalism: the Spirit of a neo-Renaissance

Past history reveals that a new spirit has always supplanted the old when society has begun to disintegrate and the emergence of a new society has begun. For example, when feudalistic agricultural society, which had lasted for many hundreds of years, began to break up, and industrial society began to emerge, the Renaissance spirit became the new spirit of the times.

The New Spirit of the Times

What will be the spirit of the times in the information society? What will be the concept characterizing the thought of the times? It will be *globalism*. This will actually signify a neo-renaissance in the sense that it will emerge in the passing of the old society, which will give place to the new society (industrial society to information society), the chief aim of which will be the liberation of the human spirit. The character and content of this neo-renaissance will differ fundamentally from the earlier Renaissance in two ways.

The first characteristic of globalism is what we may call *spaceship thought*[1]. If we term the Renaissance an era of explosion, the neo-renaissance is an era of implosion. The old territorial frontiers that divide mankind are breaking down, a fact graphically illustrated by the shortage of natural resources. Particularly noteworthy is the growing shortage of fossil fuels and metals, such as petroleum, copper, lead etc. We recognize this as a real problem, the emergence of an entirely new situation since the industrial revolution, in which, even if industrial productivity is raised and material wants are increased, industrial production will sooner or later level off because of shortages of raw materials.

Man succeeded in making a soft-landing on the moon, but tens of thousands of people cannot be settled on the moon, as was possible in settling the American continent. Mankind has no alternative but for the 4,000 million earth people to fulfil their destiny on this closed planet. Herein we find the historical background to spaceship thought.

The second characteristic is *the idea of symbiosis*. This is the concept of peaceful symbiosis, the symbiosis of mankind and nature, a new thought of our time, prevailing over liberalism and individualism. The historical background to this is the development of the ultimate sciences, along with the pollution problems that have arisen. The first ultimate science is nuclear science. The development of nuclear energy, which threatened to exterminate mankind in a thermonuclear war, led to the peaceful coexistence system between the United States and the Soviet Union predominating over ideological differences, and placing curbs on nuclear weapons. The subsequent developments of ultimate biological sciences, including the conversion of chromosomes and artificial impregnation, make it urgent that a new view of ethics be established, particularly affecting the ultimate problem of life.

The essence of the pollution problems is in the enormous industrial productivity made possible by the application of science and technology to mass production, and the emergence of a society of high mass consumption. The consequent discharge of enormous quantities of industrial and consumption wastes is seriously harming nature and causing environmental disruption, so that now a serious danger threatens the normal life and health of human beings. The appearance of pollution and the damage to nature led to the new science of ecology, from which the idea of the symbiosis of nature and man arises.

The third characteristic is the concept of *global information space* (GIS). As distinct from conventional geographical space, this means space connected by information networks. It is space without regional boundaries. When this information space is expanded to global proportions, it will be global information space, formed on the basis of a global information infrastructure of communication lines, communication satellites and linked-up computers.

Information has *no national boundaries*. When global information space is formed, worldwide communication activities among citizens that cross all national boundaries will be set in motion, and as mutual exchanges of information expand, so will mutual understanding deepen, touching problems that lie outside the boundaries of nations and states (e.g., the population explosion and energy problems), making it possible to deal with these problems from the global standpoint. When this happens, the spirit of globalism, prevailing over conflicting national interests and differences, will become broadly and deeply rooted in the minds of the people.

Notes

1 K. E. Boulding, *The Economy of Love and Fear: A Preface to Grants Economics*, Wadsworth: California, 1973.

5

Time-value: a New Concept of Value

What is Time-value?

This new concept of value will come with the future information society, for time-value will be the major determinant of modes of action.

'Time-value' is *the value which man creates in the purposeful use of future time.*[1] Put in more picturesque terms, *man designs a goal on the invisible canvas of his future, and goes on to attain it.* As we have said, time, by which we mean the measurement of the passage of time, is an intangible, abstract concept. But if conceived of as a person's lifetime, time used for the satisfaction of wants, time itself creates value.

The development of information productivity through computer-communications technology has given rise to a new basic concept of time-value to replace material values. The relations between information productivity and time-value are set out below, showing how the development of information productivity produces time-value.

The first point is *the increased effectiveness of purposeful action.* Computer-communications technology makes it possible to mass-produce foreseeing, logical and action-selective information. As a result, the effectiveness of purposeful action is greatly increased.

This change in the pattern of action necessarily tends to place more importance on time-value, or the effective utilization of time.

The second point is *the importance attached to time as a necessary ingredient of a compound process.* Knowledge-oriented information created by computer-communications technology has the character of compound and normative information. Such information eliminates much of the limitation on the scale and time of purposeful action, and so the process becomes an important factor in producing time-value.

'Process' means here the *interaction of a purposeful subject on the field or in the compound space in which the subject acts.*

The third point is *the increase of free time.* Computer-communications technology greatly increases automation functions in material production. The replacement of man's feedback functions in material production promotes automation in material production, and liberates man from the restraint on time for material production, and so increases free time.

For these three reasons, a substantial improvement in information productivity through computer-communications technology makes it possible to create new time-value to replace the conventional material value.

The Framework of Time-value

The value, as seen from the standpoint of a value system, comes within a triple framework: subject of action, field and process. The reason why these difficult concepts – subject of action, field and process – are introduced here is that the concept of time-value becomes clear only when seen within this triple framework.

As for 'the subject of action', i.e. *the subject which works on the field with objective-consciousness*, this may be any individual, group of individuals, or organization engaging in social action with deliberate purpose; an individual, enterprise, nonprofit organization, local autonomous entity, government, state, consumer movement, or group of people engaged in a citizens' movement.

As for the concept of 'field', this is *the space, with concrete content, within which the subject of action acts with conscious purpose.* However, the field in this case is not an objectively pre-existing field, but is necessarily related to the working of the subject of action and also to reality.

Another new concept of field is the field of *information space.* This is the field provided within the new space, which had never previously existed and which is connected with the networks of information. This field of information space is characterized by two features : 1) it does *not have boundaries* like a territorial field and 2) in this field, *elements related by objective-oriented action are related to each other through information networks.* The concept of field in information-oriented society will be represented more and more by this concept of information space (see figure 5.1).

As for the third concept, 'process', this is *the development in time of a situation created artificially by the interaction between the purposeful action on the field of the subject of action and the reaction of the field to it*, and it is regarded as the dynamic process of a system comprising both the subject of action and the field. By the artificially created situation is meant a situation purposefully created by the objective-oriented action of the subject of action so that the objective may be attained.

Further, this process is characterized by the fact that it is finally completed. This is because the subject of action puts an end to the action when the purpose is either attained or given up.

A. Field of geographical space   B. Field of information space

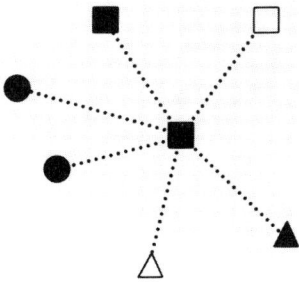

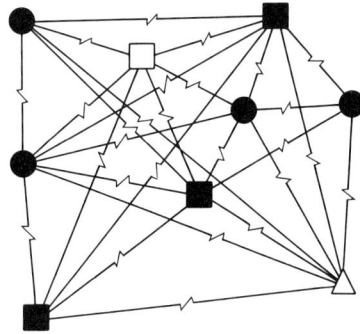

Figure 5.1 Two Types of Field.

The result of any achievement produced in the triple framework – the subject of action, field and the process – is precisely time-value, which is measured according to the degree and quality of the results achieved. Further, time-value may be the situation itself in which the process ends, or it may be the sum total of the value produced during the process, depending on the value judgement of the subject of action.

In this outline of the framework of time-value, the relationship between subject, field and process may become clearer by looking at figure 5.2. The subject of action acts on the field. Through this action, Field A, changing, shifts into Field B, which in turn, changing, shifts into Field C. The situational relationship, including both the subject of action and the field, goes on changing, shifting and

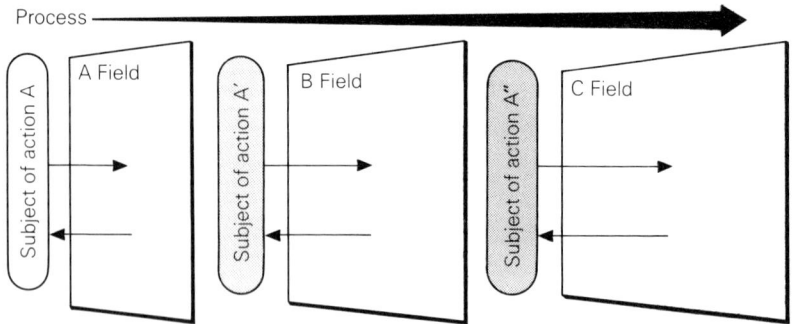

Figure 5.2 Value System of Time-value.

developing continuously. Time-value is produced in this changing and developing process.

This time-value is on a higher plane in human life than material values as the basic value of economic activity. This is because time-value corresponds to the satisfaction of human and intellectual wants, whereas material value corresponds to the satisfaction of physiological and material wants. If we define physiological and instructive wants as primary, the desire to satisfy oneself through purposeful action may be called a secondary want.

In conclusion, it may be said that the increase in productivity of knowledge-oriented information is the only motivating force that can directly heighten time-value.

Notes

1 Y. Masuda, 'Triple Concept of Information Economics', Proceedings of the Second International Conference on Computer Communication, Stockholm, 1974.

6

The Information Utility: Societal Symbol of the Information Society

While industrial society is a society formed and developed around the production of material values, the information society will be formed and developed around the production of information values. In industrial society the modern factory is a large machinery production facility which has the central function in the production of goods. This modern factory, the present societal symbol, supplanted the farm, which had been the main base of production in agricultural society. In the information society of the future the *information utility* will become the base of production of information values, and thus could appropriately be called the societal symbol of the information society.[1]

What is an Information Utility?

An 'information utility' is *an information infrastructure* consisting of public information processing and service facilities that combine computer and communication networks. From these facilities *anyone, anywhere, at any time will be able easily, quickly and inexpensively to get any information which one requires.*

The following four requirements will be essential to an information utility.

1   Central facilities equipped with large-scale computers capable of simultaneous parallel processing, connected with large-capacity memory devices, a large number of program packages and extensive data bases. Such facilities will be able to do information

processing and provide services for a large number of users at the same time.

2 Such information process and service facilities will be provided for the use of the general public, for which purpose the computers of the central facilities will be connected by means of a communications circuit directly to personal computers in businesses, schools and homes.

3 Any user will be able to call the local centre of the information utility to have data processed, or one will even be able to process the necessary data for oneself.

4 The cost of using the services of the centre must be low to enable the general public to use the centre readily for day-to-day needs.

There is a system already in existence which partly meets these conditions. It can be seen in the commercial time-sharing service and the utilization of computer networks operating between universities. But these systems are restricted to certain social strata (large enterprises and universities), and the costs are high. It was once thought that it would be several decades before information utilities could become available to meet all these conditions fully, but the development of ICs (integrated circuits) in the early 1970s ushered in the era of fully fledged information utilities.

There are at present two main currents in the world to support this claim. One is the experiments on community information systems, such as the Tama CCIS and Hi-Ovis of Japan and Sweden's Project TERESE, and the other is the development and practical application of the *videodata systems*, including those of the United Kingdom (Prestel), West Germany, the United States, Canada (TELIDON) and Japan (CAPTAINS).

Videodata systems allow users telephone access on demand to information stored in computers, which is then displayed for them on modified television sets. The system uses span publishing, marketing, in-house communication, public service information, education and entertainment.

These systems have the ideal features required for an information utility, including (1) they supply diversified information, (2) they require only a small investment ($300–$500), (3) service charges are as low as telephone charges (10–20 cents per operation), (4) they are easy to operate (select push-button system). These factors open up the possibility of tens of millions of sets being installed throughout the world, even by the 1980s. The advent of fully fledged information utilities can be considered now to be on the horizon.

Can we now project the picture of a future information utility as the societal symbol of the information society?

A desirable and feasible future information utility will represent the integration of three concepts – (1) information infrastructure, (2) joint production and shared utilization and (3) citizen participation.

Formation of an Information Infrastructure

The first fundamental of the information utility as we conceive it is that *it will take the form of an information infrastructure.*

There are two reasons for saying this. First, the information utility will be required to have all that is essential to the other parts of the infrastructure, including electricity, water, railroads, etc. The information utility will (1) become indispensable to the support, development and maintenance of socio-economic activity, (2) require massive investments in equipment and facilities, (3) be linked in a regional and/or nationwide network. Information utilities will share all three conditions with the existing components of the infrastructure.

Second, the information utility by its very nature will be for the use and benefit of the public, its service being of a unique character, that is *self-multiplication*. Information, unlike material goods, has four inherent properties that have made self-multiplication possible.

1 It is *not consumable* – goods are consumed in being used, but information remains however much it is used.
2 It is *non-transferable* – in the transfer of goods from A to B, they are physically moved from A to B, but in the transfer of information it remains with A.
3 *Indivisible* – materials such as electricity and water are divided for use, but information can be used only as 'a set'.
4 *Accumulative* – the accumulation of goods is by their non use, but information cannot be consumed or transferred, so it is accumulated to be used repeatedly. The quality of information is raised by adding new information to what has already been accumulated.

Synergetic Production and Shared Utilization of Information

The second fundamental is that both *the production and utilization will be a combined operation*. This means that the production structure of information is *self-multiplying*.

'Self-multiplication' does not mean the successive production of new information, but *the utility's continuous expansion in the production of information, both in quantity accumulation and qualitative improvement*. All information produced by the information utility is cumulative, with new information constantly being added. In simpler terms, *the accumulation of information leads to further accumulation of information which in turn means still further accumulation of information over time and space*. One can say that information utilities are systems that demonstrate to the maximum the self-multiplication function of information by man-machine methods.

There will be four stages of development by which the man-machine self-multiplication production systems of information utilities will come to maturation, viz., 1) public services, 2) user production, 3) shared utilization, 4) synergetic production and utilization.

Public Services Stage   This is the stage at which the information utility provides information processing and services for the public. Many kinds of programs and data bases are prepared by the utilities in advance, and users utilize the information service within the limits of this preparatory stage. (Videodata systems belong in this stage.)

User-production Stage   At this stage, the user of the information utility produces information. The user collects data, makes the program, and makes use of the information utility to produce the user's own required information. There are four factors to promote user-production of information, the first being the awareness of the general public that ones own information can be produced for oneself. The second will be the development of sophisticated language programs in the conversational mode. The third will be the development of various packaged program modules, and the fourth will be the preparation of data bases to suit many different fields. These factors will enable the general public to become aware both of their ability to produce and of the value of producing their own information.

Shared Utilization Stage At this stage, the information utility makes possible the shared use of information produced by individual users. As the production of separate information by individuals reaches a certain point, the data and programs are registered and become available to third parties, and the self-multiplication process and shared utilization interact to produce a geometric effect.

Synergetic Production and Shared Utilization Stage The synergetic production of information by a group belongs to this stage. The shared use of information by the voluntary registration of programs and data developed by individuals develops into voluntary synergetic production and shared utilization of information by groups. It is evident that several people working together in collecting and processing data will be more efficient. There will be frequent need for complex programs that will only be possible when several people work together in the development and utilization of the product. This is obviously so in the development and shared utilization of complex technology and programming to resolve the more complex problems.

Synergetic production and shared utilization of information will be the most developed form of information production by the self-multiplying operation of the information utility.

Three Types of Information Utility

The third fundamental is *citizen participation*. Three possible types of management can be envisaged for the information utility: the business, government managed and the citizen managed types.

In the long run, the citizen-oriented mixed type will probably become predominant because citizen participation will be essential to the management of information utilities. This becomes clear if one looks at the shape that these three types of management can assume, and then notes, from a macro point of view, the socio-economic merits and demerits of each.

Business Type

The business type of information utility will be privately capitalized, and its functions will be exercised on a wholly commercial basis by free competition of private enterprise. Operations will be based on the income to be derived from information processing and services. The major types of services will be provision of information related to everyday convenience in the lives of the general public (news requests, information on shopping, etc.), or concerned with various sorts of mental exercises or recreation (spaceship games).

Chief merits of the business type are that efficiency in management will be essential and the services thorough. Negative values would result from excessive commercialism resulting in information services encouraging mental laziness and stagnation by the emphasis placed on convenience, accompanied by aggressive advertising.

Government Management Type

The capital required for the governmentally managed information utility would be provided from the national budget, the goal of which would be to increase the well-being of the people as a whole. The national government would operate the utilities, with the dual support of taxes and revenue from the utility rates in payment for services. The services would include all forms of public relations, information about government policy, statistics, information to serve the public interest (weather, pollution, transportation, etc.), and information services of a social welfare nature, such as education and medical care.

The chief merits would be the low rates for usage and the requirement that such utilities exist for the public good. Negative values would derive from the inefficiency associated with bureaucratic organization and the danger of increased governmental control over society.

Citizen Management Type

Here, a third type of capital is civil, differing from private capital, raised by citizens themselves. Operations would be completely under *the autonomous management of the citizens*, with the operational base consisting of *funds raised by citizens*, from usage fees and *voluntary contributions* (these would include money, mental labour and programming).

Both the processing and supplying of information will be essentially in the form of joint production and shared utilization by the citizens themselves, with types of information related to problem solving, opportunity development for individuals, groups and even society as a whole.

Here, the merits will be maximal voluntary participation by citizens, allowing the individual to obtain the information needed. It becomes much easier to arrive at a solution and the direction for joint action to solve common social problems. A demerit is that in capital formation, technology and organization, this type is inferior to the previous two because functionally it would depend to a very great extent on the voluntary contributions of citizens, which would be difficult to coordinate.

The three types of management for the information utility (summarized in table 6.1) have here been stereotyped and somewhat exaggerated as to the characteristics of each. Considered realistically, the information utility of the future is likely to be a combination of two or three types.

Table 6.1 Three types of information utilities

	Business type	Government managed type	Citizen managed type
Management goal	Profit	Welfare	Information accumulation
Type of capital	Private capital	Government capital	Civil capital
Type of management	Private	Government	Autonomous
Form of production	Time sharing services	Data banks	Synergetic production
Area of service	Daily convenience, leisure	Medical care, education	Problem solving, opportunity developments
Operations base	Sales revenue	Usage rates, taxes	Voluntary contributions, usage rates
Price system	Free price system	Public utility rate system	Income standard system
Merits	Efficient, good service	Operated in the public interest, inexpensive	Autonomous, creative
Demerits	Commercialism, mental degeneration	Danger of control, inefficiency	Weak, unstructured operations base

Citizen Participation is Essential

Whatever the combination, citizen participation is an essential condition, the most desirable form of which would be mixed and citizen-oriented. The main reason is that (1) only by citizen participation in the management of information utilities will the self-multiplicative production effect of information be expanded; (2) autonomous group decision making by ordinary citizens will be promoted; and (3) the dangerous tendency toward a centralized administrative society will be prevented.

The Macro-cumulative Effect of Information

The first reason why information utilities will tend toward citizen management is that this type of information utility, more than any other, will facilitate the macro-cumulative effect of information. I have already mentioned that self-multiplication of information production gives the information utility the nature of an information infrastructure. When viewed from the perspective of the national economy, the self-multiplication can be called *the macro-cumulative effect* of information; the effect being in sharp contrast to *the mass production effect* of goods produced in the modern factories of industrial society.

In the production of goods, the expansion in manufacturing that follows a big increase in production equipment has a great mass production effect. That is to say, the greater the investment in capital equipment, the more productive power increases and production costs decrease. This decrease in costs expands the market and encourages further profits and further accumulation of capital. This mass production effect of goods is, from the enterprise's point of view, the multiplying effect of capital, in the sense that any accumulation of capital results in further accumulation of capital.

The self-multiplication of private capital has been a fundamental cause of the formation and expansion of modern manufacturing industries as a whole.

In the case of information also, expansion in the scale of production cannot be ignored, but here the cumulative effect is more important.

The most important point in the production of information from a macro standpoint is *the self-multiplication of information value* itself –

how to accumulate information and how to continue the further accumulation of information by adding new information to what has already been accumulated.

The information utility is not used simply by a limited group of users; it is widely used in the public interest by people in general. Moreover, it is the general public themselves that operate the information utility freely. Having the information utility take the form of synergetic production and shared use will raise the macro-cumulative effect of information utilities to the highest level. The citizen management type that is oriented toward voluntary synergistic production and shared use of information by citizens themselves is the form of management of the information utility that will have the greatest macro-cumulative effect, rather than the business type that aims to increase profits through the self-multiplication of capital, or the government managed type that prevents citizens from using the information utility freely.

Autonomous Group Decision-making

The second ground for confidence is that *this type makes autonomous group decision-making possible*, with the aim of solving complex socio-economic problems through autonomous decisions by the citizens themselves.

The inadequacy of governmental compulsion and monetary compensation in solving the problems of human existence will stimulate these possibilities even further. The future information society will be a society in which autonomous decision-making will be the most basic human right. The causes of problems that will arise in the future will be very complex and interrelated, and in complex opposition to these will be the individual group interests of the citizens. In solving such problems, *mutual understanding* and *voluntary cooperation* of each citizen in selecting the action that corresponds to one's own situation will be essential.

Avoiding a Controlled Society

The third reason is that only through this type can *the information society avoid the dangers of a controlled society*. For the information society to become an ideal society of voluntary decision-making, and to avoid the ultimate fearful Orwellian automated state, will depend on the form of management adopted for the information utility.

If information utilities were to be completely dominated by a despotic state organization, the information society would be the ultimate controlled society, in which the abuses would by far exceed the alienation of man in the present industrial society or the abuse of human rights under dictatorships. This could occur because, in information utilities, both public and personal information concerning each individual is filed and accumulated, and, in addition, information about one's major activities in society would be added constantly to this file.

But if the information utilities are completely entrusted to the voluntary management of the citizenry, and if the personal information of individuals is completely protected and used to improve the private life of each individual and the quality of one's social activities, then the information utility will be of immeasurable benefit to all citizens. For example, the information utility will not simply provide the individual citizen with information that is useful in solving everyday problems (illness, work, learning, housing, etc.), but will also contribute greatly to maintaining the individual's life in a healthy active state, by combining this sort of social information with personal data about each individual.

Looked at in this way, the information utility is a socio-economic institution which concentrates the ultimate in scientific technology. In this sense also *the autonomous management by citizens* of the information utility is an essential prerequisite for the ideal information society.

Vision of a Global Information Utility

The information utility will extend to an international scale; it will reach a substantial level of development, and then will become a GIU (global information utility).[2]

The concept of a GIU projects a global information infrastructure using a combination of computers, communication networks and satellites. Its basic feature would be that any ordinary citizen in the world could obtain all necessary information readily, quickly and at low cost, at any time and place in the world.

A GIU would operate on the following minimum requirements:

1 Several global information switching centres (GISCs) throughout the world, each of which would be connected to several scores of sub-global information utilities (sub-GIUs), would be required. Each sub-GIU would be equipped with a number of large capacity computers capable of on-line real-time processing of information.

2 GISCs would be mutually connected by satellites so that users could utilize not only sub-GIUs in their areas but also sub-GIUs in any place in the world through the GISCs.

3 Fees for GIUs services would need to be low enough for ordinary citizens in any country of the world to be able to use the facility on a day-by-day basis.

4 The basic computer language for GIUs should be an internationally standardized language, whereas input–output languages for individuals in different countries would have to be their native languages. For this purpose, each GIU would have to be equipped with an automatic translation system.

When GIUs come into practical application, it will be possible for people anywhere in the world, for instance, to call to their aid such services as CAI-oriented self-education systems, library and other information services, world news services, comparative studies of incomes and pensions with other countries, planning overseas travel for oneself, and competitive mental games according to time differences between different parts of the world. Imagine an international TV game contest. Thousands of different TV games will be possible, with contests among international mixed teams; GIU prizes can be awarded for the invention of new and exciting games. A computer art contest could be very colourful and artistic.

The following ideas would be of the greatest significance to mankind. Take, for example, the accumulation of data on air-pollution at GIUs from time to time, gathered from many thousands of places in the world; this would become a global air-pollution information and correction system. The enormous volume of air-pollution information thus collected by GIUs would be at hardly any cost, by the voluntary cooperation of citizens in many countries.

Of more revolutionary significance would be *a global voting system*, by which hundreds of millions of people in the world would be able to participate in making decisions on global problems – such as nuclear power generation and SSTs – which could have an unpredictable effect on the whole human race.

If such a vision of GIUs were to become a reality, it would have an incalculable impact on human society.

1 As such exchanges of information among ordinary citizens in different countries of the world become a reality, a supra-national GIU system, overriding national interests, would be established as a result of increased cohesion of citizens.

2 The establishment of a global CAI education system would enable 90 per cent or more of the world's population to become literate, and a world language, distinct from Esperanto, would ultimately be developed.

3 The functioning of a global medical care and pollution prevention system would eliminate leprosy, malaria and other endemic diseases, and lengthen the average life span of humans to 90 years or more. At the same time, birth control would be effectively practised so that the total population of the world could be stabilized at 5,000 million or so.

4 The south–north gaps in wealth and cultural levels would be narrowed, and as values were diversified and individual and autonomous group activities were encouraged, a widened range of creative cultures would flourish.

5 A new society, with new economic principles, would come about, consistent with the basic characteristics of GIUs – the global joint creation and utilization of information. Thus, the transformation from the present individualistic principle of free competition to the principle of synergic activity among independent individuals cooperating functionally for a common objective would eventuate; human society formed on the principle of synergetic cooperation would mean a global society based on mutual assistance.

If what we have envisaged above is realized, GIUs can be expected to develop by geometric progression in the coming decade. This calls for preparations as listed below to be made as soon as possible to promote the formation of GIUs to meet these international needs.

1 Conclusion of *an international treaty on the joint control of communication satellites*.

2 Establishment of *an international information development organ* to promote the establishment of GIUs.

3 Formulation of *concrete medical care and education projects for developing countries*.

4 Promotion of *standardization of hardware and software related to GIUs*.

Notes

1 Y. Masuda, 'Future Perspectives for Information Utility', Proceedings of the Fourth International Conference on Computer Communication, Kyoto, Japan, 1978.
2 Y. Masuda, 'A New Era of Global Information Utility', Proceedings of Eurocomp 78, London, 1978.

7

A Synergetic Economic System: an Information Axis Economy

In considering the economic structure in relation to the system of the information society, the information society will be one that develops around the production of information values, as I have said, and will therefore differ fundamentally from the agricultural and industrial societies of the past, which developed around the production of material values. More precisely, the term 'information society' refers to an economy in which (1) *information is the core of society's economic needs*; (2) *the economy, and society itself, grow and develop around this core, the production and use of information values*; and (3) *the importance of information as an economic product exceeds goods, energy and services*. This economic structure could be called an 'information axis economy'.[1]

On the supposition that the information axis economy centring on the production of information comes into being, what changes will it bring about in the system and structure of the economy?

Change to an Information-led Type of Industrial Structure

First, there will be *a change from an industrial structure centring around goods, energy and services to an information-led type of industrial structure*, a change that will pass through three stages of development.

The Appearance of Information-related Industries

The first stage will be *the formation of information-related industries*. In an information society the information-related industries will become the leading industries, which will develop to the point of formation of *quaternary industries* as a new classification.

The concept of 'quaternary industries' is necessary as a classification, in that a clear line of demarcation must be drawn between service industries and information-related industries. There is a general tendency to characterize the major change in the structure of post-industrial society as an expansion of tertiary industries (i.e. service industries) with their importance exceeding that of the secondary (manufacturing) industries. In this context, information-related industries could be classified as tertiary industries only because they are non-goods-producing industries. But it is highly probable that information-related industries would develop beyond the service industries in an information society. It is reasonable therefore to distinguish information-related industries from service industries, and classify them as quaternary industries, to provide a clear concept of the industrial structure of an information society.

What would be the composition of the quaternary industries, as leading industries of the future? Quaternary industries can be divided broadly into four main industrial groups: (1) information industries; (2) knowledge industries; (3) arts industries; and (4) ethics industries. Of these four, the information and the knowledge industries will become the key industries of the future. Figure 7.1 gives a breakdown of all information-related industries.

The first 'information industries' will primarily be industries that produce, process and service cognitive information, or produce and sell related equipment. Here, the information industries are restricted to cognitive information because a separate industrial group, the arts industries, will be established for affective information. The core of the information industries will comprise the information machinery industries, including the computer and peripheral equipment industries, and LSIs (large scale integrated circuits) and micro processors. Such information machinery industries will probably displace the automobile industry to take first place as the largest manufacturing industry.

Strictly speaking, one could classify these information machinery industries as secondary industries, but information-related industries are here given a broader meaning and come under the term 'quaternary industries'.

In parallel with this, the industries concerned with information processing services and systems development, software, TSS services

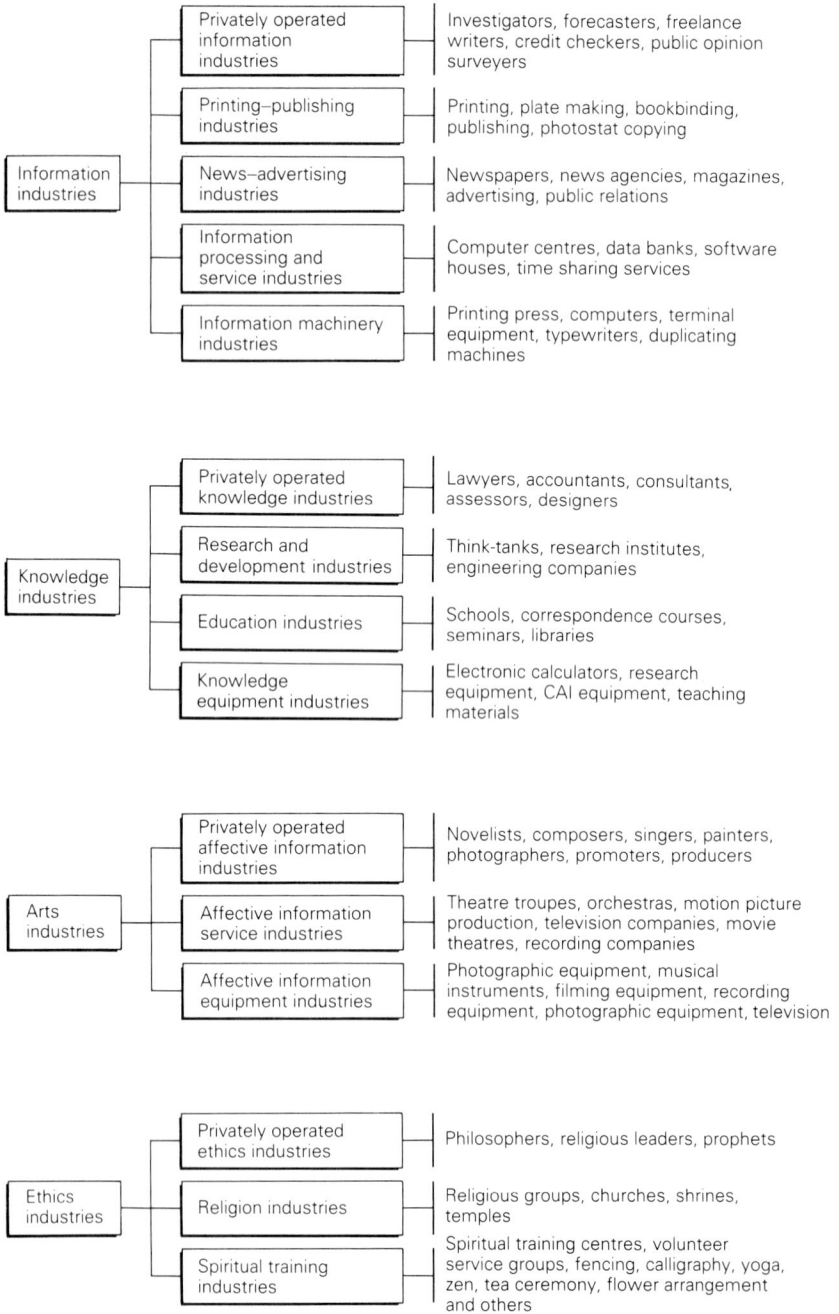

Information industries	Privately operated information industries	Investigators, forecasters, freelance writers, credit checkers, public opinion surveyers
	Printing–publishing industries	Printing, plate making, bookbinding, publishing, photostat copying
	News–advertising industries	Newspapers, news agencies, magazines, advertising, public relations
	Information processing and service industries	Computer centres, data banks, software houses, time sharing services
	Information machinery industries	Printing press, computers, terminal equipment, typewriters, duplicating machines

Knowledge industries	Privately operated knowledge industries	Lawyers, accountants, consultants, assessors, designers
	Research and development industries	Think-tanks, research institutes, engineering companies
	Education industries	Schools, correspondence courses, seminars, libraries
	Knowledge equipment industries	Electronic calculators, research equipment, CAI equipment, teaching materials

Arts industries	Privately operated affective information industries	Novelists, composers, singers, painters, photographers, promoters, producers
	Affective information service industries	Theatre troupes, orchestras, motion picture production, television companies, movie theatres, recording companies
	Affective information equipment industries	Photographic equipment, musical instruments, filming equipment, recording equipment, photographic equipment, television

Ethics industries	Privately operated ethics industries	Philosophers, religious leaders, prophets
	Religion industries	Religious groups, churches, shrines, temples
	Spiritual training industries	Spiritual training centres, volunteer service groups, fencing, calligraphy, yoga, zen, tea ceremony, flower arrangement and others

Figure 7.1 The Quaternary Industries (Information-Related Industries).

and computer centres, will also undergo unprecedented development. The present mass communications industry of newspapers and publishing, which dominate the information services, will probably enter an era of stagnation.

The second group, the 'knowledge industries', can be expected to develop after the information industries have come into being. The core of these will be of two types, viz., education industries and research and development industries. The education industries and the information industries together will be the pillars of the information society. The reason for this is that in the information society human values will change; material values will be superseded by time values, and greater importance will be attached to the development of new abilities and the improvement of human life. Research and development industries will greatly expand in response to the need to solve problems of resources and energy, an example of which would be resources recycling technology, or the need to improve human welfare through integrated social systems.

The third information-related industrial group is the 'arts industries', such as the mass communication industries (newspapers, publishing, etc.). These seem to have reached a peak now, and it is likely they will be declining industries. The reason is that in an information society the expansion of individual creative knowledge will flourish, and society will be able to escape from the present fad-oriented, sensory, television-dominated society of today. The future of television is probably that it will be linked with the computer and function more as a medium for cognitive information. One main use, for example, would be for citizen participation in decision-making on social problems.

The fourth group, the 'ethics industries', by contrast, will become growth industries, among which religion will form a particularly important part. In this case, while moving away from belief in the existence of a supernatural god, religion will be epochally significant, in that human life will be elevated through renewed belief in the existence and strength of humanity. In the information society, on one hand, each person will attach more importance to scientific thought, and on the other, will be humble before an absolute existence that transcends human abilities. The very basis of this humility will be the global concept, the harmony and symbiosis of man and nature on this finite planet, earth. Whether one calls this globalism a religion or calls it the awareness of a supra-human existence, I believe the future will call for religious thought and an ethical content that are new and clothed in new attire.

In the information society of the future, the industries in these four groups, together with the computer industries, the information processing and service industries, the education industries and the religion industries, will form a core group whose major function will be in the growth and development of the quaternary industries as a whole.

The Formation of Industries with Installed Information Equipment

The second stage will be marked by *the formation of industries with installed information equipment.* This means the informationalization of industries through utilization of the machinery, with information equipment forming part of it. The development of machinery and equipment capable of exercising information functions has been made possible through the invention of LSIs and micro-processors. We have already reached the stage where a wide range of machinery with installed information equipment, such as electronic calculators, electronic watches (digital or other), automatic cameras, cash dispensers, games machines, etc. is on the market. In the future, it can be expected that machines and facilities containing information equipment will come into extensive use in medicine, education, pollution control, traffic control and all industrial areas. As this process continues, various kinds of robot-operated and control machinery and equipment will certainly be created.

The Development of Systems Industries

The third stage will be *the development of systems industries.* Analyses of the industrial structure have, in the past, generally been based on a quantitative assessment setting out the ratio of constituent industries, and the shift from low to high technology industries.

But the structure of the systems industries will consist of a complex of industries formed by linking up existing industries with the information industries. This means a qualitative change in the industrial structure, one example of which is seen in automatic warehousing, which has combined warehousing with the information industries. The systems industries may range from the relatively simple, such as automatic warehousing and automatic diagnosis systems, to the highly complex, such as the health industry and the opportunity industry.

The 'health industry' will comprise a system that includes food, pharmaceuticals, medical services, sports and information, combined organically. The medical care industry of the future will not mean simply the treatment of disease. The emphasis will be on early diagnosis and the maintenance of health. Medical care, through the integration of new systems and technology, such as frozen food technology, preventive medicine, hospital automation, health diagnosis centres, and athletic clubs, can be expected to develop into a health industry that operates as a systems industry.

What we have called the 'opportunity industry' will be an integrated industry whose functions will be to open up personal possibilities for the future, and like the health industry, will be one of the systems industries that offers the possibility of personal growth.

The four major opportunity industries will be education, information, ethics and finance–insurance, each with its own function. The education industries will promote the development of individual abilities. The information industries will supply information that will enable new opportunities to be discovered and created. The ethics industries will provide behavioural standards and guide in the moulding of character, and the finance–insurance industries will provide needed capital and risk cover. The education industries will not operate on the current uniform system of school education, but will become a more diverse and dynamic system of education. Home self-learning, labour re-education and computer plazas (where people can use the computer freely) will make their appearance. A mechanical data bank of enormous range will be established as an information industry, and the accumulation of information will proceed on everything necessary for opportunity development, from analysis of individual potential and work references to the location of educational facilities.

Training facilities for the ethics industries will be provided, and community centres where values and behavioural standards, thought, religion and ethics are the focus of interest. The finance–insurance industries will be equipped to provide opportunity loans, the capital needed for the application of ability and opportunity development, and opportunity insurance as a guarantee against the risk of the loans.

The labelling of such systems industries should not be according to industrial categories used for traditional types of goods; the systems should be related to traditional categories of industries by means of a matrix. If the traditional categories of industry from primary to quaternary are placed on a horizontal axis, and systems industries are placed on the vertical axis, the relationship can be established. Table 7.1 sets out this new industrial matrix.

The primary industries, divided into agriculture and forestry, stock-farming, fishery, mining and manufacturing industries, are placed on the top portion of this industrial matrix. The more specific industries in each of these, such as mechanized agriculture and the broiler industry, are included. The industries shown in the matrix are those that bear some relation to the systems industries, which are on the horizontal axis. This matrix enables a multi-faceted quantitative and qualitative analysis of the industrial structure to be made in a way that has not been possible before.

Expansion of the Public Economy

The second change in the economic structure will be *an expansion of the public economy*. Expansion of the public economy refers to an increase in the public side of economic activities, with the emphasis on economic activity for the public benefit rather than on profits to be made.

This expansion of the public economy will occur in four ways.

Strengthening of the Infrastructure

The first form of expansion will be *the strengthening of the infrastructure*. In the current industrial society, gas, water, communications, roads, parks, schools, etc., are typical examples of the infrastructure. But in the information society many other kinds of facilities and services will become important parts of the infrastructure. The most significant, needless to say, will be the information utilities.

In the information society, information utilities will be the driving force of social development. The information utilities themselves will basically assume the nature of an infrastructure, which alone will become a decisive factor in the transformation of the economic structure from the present private economy into a public economy in the information society.

In addition, *lands may also become part of the infrastructure* because these are limited basic resources that will take on the character of public assets, which will result from the separation of land ownership from use.

Of special note as infrastructure will be *do-it-yourself facilities*. These will be public facilities where many things can be done: carpentry, woodworking, ceramics, fabric dying, weaving and even machine construction. These facilities will satisfy people's need to make things for themselves. Such places will be equipped not only with tools for handicrafts, but with highly sophisticated machinery and the needed technology for people to make things for themselves that may take several months to complete. Hobbies for leisure-time activities are becoming more and more popular even now, but the information society will greatly increase free time so it is expected that these do-it-yourself facilities will develop to an astonishing degree, in sharp contrast with the antihuman automation-based production in present-day industrial society.

Basic Material Industries Join the Public Economy

The second form of expansion will be that *basic material industries join the public enterprises*. In the future information society, steel, oil refining, petrochemicals, fertilizers, synthetic fibres, aluminium, and other basic materials industries will probably all come into the public sector, for which there are two impelling reasons.

First will be the growing threat of shortages in basic material supplies, due partly to the depletion of natural resources, changes in climate, increases in population and other factors. No matter how sophisticated technology becomes, it will not be possible for technology to produce adequate supplies of resources artificially, or for resources recycling to be total. Shortages in the supply of basic resources will inevitably become a major socio-economic factor, restricting the free economic activity of the private enterprise system.

A second reason why basic industries will become public is that in these basic materials industries automation by the use of computer technology will make very rapid progress. Already automation in industrial production has reached a fairly advanced stage, and before the beginning of the twenty-first century there will probably be

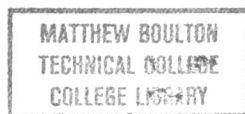

MATTHEW BOULTON
TECHNICAL COLLEGE
COLLEGE LIBRARY

Table 7.1 Matrix industrial structure

			Systems Industries										
	By products	By systems	Distribution industry	Integrated Transport industry	Housing industry	Regional development industry	Environment industry	Marine industry	Space industry	Leisure industry	Fashion industry	Health industry	Opportunity development industry
Primary Industries	Agricultural industries	Mechanized agriculture										●	
		Broiler industry										●	
		Fish hatcheries										●	
	Fisheries industries							●					
	Mining industries	Offshore oil wells						●					
		Underwater mining						●					
Secondary Industries	Light industries	Manufactured foods (margarine, etc.)							●				
		Frozen foods										●	
		Clothing								●	●	●	
		Cosmetics									●	●	
		Medical supplies										●	
	Heavy, chemical industries	Artificial organs										●	
		Medical engineering equipment										●	
		Anti air-pollution devices				●	●					●	
		Construction equipment				●	●					●	
		Ambulances		●	●	●						●	
		Disaster prevention devices										●	
		Rack style warehouses			●				●				
		Deep sea submarines						●					
		Space devices					●		●				
		Communications satellites					●	●	●				
		Traffic signal equipment				●							

Construction industries		Prefabricated houses
		Mobile homes
		Hospitals
		Highways
		Land development
Tertiary Industries	Utility (light, heat and power)	Atomic power
		Solar energy
		Natural gas
	Traditional public utilities	Regional heating and air-conditioning facilities
		Monorail
	Freight industries	Container transport
	Communications	Data communications
	Commerce	Supermarkets
		Distribution centres
	Warehousing	Warehouses
	Finance and insurance	Opportunity loans
		Housing loans
	Personal services	Medical care
		Human docks
		Athletic clubs
Quaternary Industries	Information industries	Data banks
		Computer Centres
		Software
		Computers
		Terminal equipment
	Knowledge industries	Think tank
		Consultants
		Audiovisual teaching equipment
		CAI equipment
		Colour television
	Arts industries	Movies
		Plays
		Records
	Ethics industries	Spiritual training centres
		Religious groups
		Volunteer service groups

massive and complete automation in basic industries, where the merits of scale are great, such as in the steel, petroleum, petro-chemicals, cement and electric power industries. The increasing shortages in the supply of resources, combined with the further expansion of productive power by applied automation will result in increased contradictions between the profit principle of monopolistic private capital and the public interest.

Expansion of Social Consumption

The third change in the public economy will be *the expansion of social consumption*. The dominant form of consumption in industrial society at present is individual, as evidenced by food, housing, automobiles, etc., all of which are commodities of personal consumption. The examples of social consumption are parks, roads, schools, hospitals, etc., but individual consumption carries far greater weight than social consumption.

In the information society, however, social consumption will constitute a far larger portion of total consumption than individual consumption, influenced by two factors. The first will be that individual consumption will reach saturation point, with increased social disutility. In developed countries, the share of individual consumption is rapidly decreasing. And social disutility, which includes air pollution, urban congestion and the destruction of nature, is increasing in inverse proportion to the increase in material consumption. As these conditions increase, people will attach more importance to social utility, thus imposing restraints on individual consumption and enhancing the tendency for an increase in social consumption.

The second factor to encourage social consumption is that the basic characteristic of computer information is as a service offered by a public utility. In the information society this computer information will be used extensively by people in general through the information utility. There will also be a large number of social information network systems, such as the medical care information system and the education information system, that will be essential to the maintenance of health and the development of capabilities.

The Shift to a Synergetic Economic System

We now consider *the shift from a free economic system to a synergetic economic system* in the information society, the total transformation of the economic system itself. This will be the final form, comprehensively encompassing the transformation of the economic structure so far outlined.

For the most part, the current economic system in industrial society has tended to be a liberalistic economic system characterized by (1) free competition of private enterprise, (2) pursuit of profits, (3) commodity production, (4) supply – demand as the determinant of prices. In the future information society, however, this liberalistic economy will be transformed into a new economic system, a synergetic economy, which is *an economic system based on synergism*. The three aspects of this are as follows.

Synergetic Production and Shared Utilization

First will be the transformation from a commodity economy to *a synergetic production and shared utilization economy*. Now the production of goods is at the core of economic activity, with production and consumption wholly separated. But future society will mean synergetic production and shared utilization, as constituting the primary economic system. The realization of this will be encouraged in two ways, one of which will be the development of information utilities. In the information society, information utilities will form the core of economic development. The unique production system of the information utilities will be structurally quite unlike factory production. As has been said, the production of information by the information utilities will differ structurally from the production of material goods by factories, in that it will be man–machine based production of information, characterized by self-multiplication. Information utilities will not merely provide extensive information processing and service to the general public. The information utilities will be used by the people themselves to produce the information they require. In addition, the programs produced by the people and the data they have collected will be available for shared utilization by all other persons. When this occurs, information utilities will advance to the synergetic production and shared utilization of information accumulated by the citizens. This unique production structure of the information utilities will not determine merely the production structure for information goods; it will broadly determine the structure of the consumption and distribution of information goods. Producers will also be users, and in this way the economic goods produced will be shared and utilized. And because the information utilities will be the axial institutions in the economy of the information society, the joint production and shared utilization of information goods will greatly influence the economic structure as a whole.

One more thing to be emphasized is that the people will voluntarily participate in the synergetic construction of public facilities. In the information society, it will become quite normal for the synergetic labour of citizens to construct public facilities such as homes for the elderly, parks, roads and schools. Of course, national and local governments will provide part of the materials and funds, but the main characteristic of this kind of construction will be that the share of funds and physical and mental labour voluntarily contributed by the citizens themselves will be greater.

Voluntary Synergy to Achieve a Shared Economic Goal

The second aspect is that, corresponding to free competition, there will be voluntary synergy. This refers to individual economic subjects carrying on economic activities synergetically in order to achieve a shared economic goal.

In industrial society, private enterprises carry on business activities freely. The result has been that this free competition has meant the development of the national economy as a whole, and provided national economic welfare. This free competition in the micro enterprise economy has functioned effectively, without conflicting, on the whole, with the orderless macro national economy, because the law of price, Adam Smith's Invisible Hand, has guided and adjusted business activity. Behind Adam Smith's law of price, however, was the tacit economic assumption that resources are limitless, and if demand expands, the production of goods will go on expanding indefinitely.

But this economic assumption is now proving to be invalid; Smith's Invisible Hand is not functioning as effectively as in the past, because we have begun to recognize that resources are finite. The gigantic productive economic subjects have to give priority to carrying on economic activity to reach shared economic goals, which, in one sense, means voluntary synergy corresponding with free competition.

Autonomous Restraint of Consumption

The third aspect is *the autonomous restraint of consumption by the people*. One economic principle of industrial society is the raising of consumption levels by mass production and mass consumption. But in the information society, autonomous restraints on the consumption of goods will apply to ensure stabilized development of the economy. The economic ethics of ordinary citizens will require that the limited natural resources must be used efficiently, and inflation prevented. The idea that the problems of shortages of natural resources and inflation will be resolved not by law imposed from above, but through voluntary restraints presents a new economic concept, worthy of the information society, and it is the concept and idea of the synergetic economy that lies at the base of this system.

Increased Management and Capital Participation

The fourth aspect will involve *an increase in management and capital participation*. The tendency for labour and the general public to participate in enterprises has already become a historical fact, and in the information society this tendency will certainly increase. The private side of private enterprise will decline, and the social side will increase as the public nature of economic activity expands from management participation to capital participation by labour and the general public. As this tendency progresses, there will be a change from authoritarian synergy to functional synergy in economic units.

The synergetic relationships in economic groups in the existing economy are relationships of authoritarian synergy between the owner of capital, who has the right of management, and the employed workers. The lower stratum has followed directions from above. But economic groups of the future will move toward an economic community of people who *participate voluntarily and share the same goal*. The synergetic relations that come into being will not be authoritarian but purely functional, which will probably come about in stages by various methods and means; there will be management and capital participation by the general public, as well as *autonomous management*. What can be said clearly is that the managerial class in the information society will not be a privileged class backed by monopoly ownership of capital and therefore the right of management; it will be *a functional class that has the job of management.*

This provides the general outline of the fundamental characteristics of the synergetic economic system. The synergetic economy will not suddenly replace the existing economy in the developing information society; rather, elements of the liberalistic economy will continue for a long time. For example, the pursuit of profits and the free prices and free markets of commodity production will not just suddenly disappear. I am referring here to a gradual shift from the present economy to a synergetic economy in the sense that the hub of the economy in the information society will be a synergetic economic system, with the trinity of *contribution motive, voluntary synergy and synergetic production with shared utilization.*

Notes

1 Y. Masuda, *Information Economics*, Tokyo, Sangyo Noritsu University Press, 1976.

8

Participatory Democracy: Policy Decisions by Citizens

I want to take up the question now of a possible political system in the information society. If I may set out my conclusion first, I would say that the political system in the information society must be in the nature of *participatory democracy*. By this I mean a form of government in which policy decisions both for the state and for local self-government bodies will be made through the participation of ordinary citizens. The present political system is a parliamentary democracy in which the people elect representatives by vote, and the people participate only indirectly in decision-making in the central government or local self-government entities, with political actions in the hands of the people's representatives. In other words, it acts as indirect democracy by means of indirect participation.

The Call for Participatory Democracy

The first reason why the political system in the information society will have to be changed from parliamentary democracy to participatory democracy is that the *behavioural pattern of ordinary citizens will change*. They will be even less satisfied with mere material wants than they are now: *their chief desire will be for self-realization*. The satisfaction of material needs follows the process of production, distribution and consumption of material goods, while the quantitative and qualitative improvement of the people's material needs results from the increased capacity for material production and a better distribution of profits between capital and labour. In the long-range view, material production capacity grows at a much faster rate than the distribution in wages paid for labour.

An extreme example of this would be when material productivity of a given country grows tenfold in half a century, whereas the absolute value of workers' earnings increases only fivefold, the distribution rate for labour standing at one-half. The question we must ask is, what economic system can raise material productivity best to develop the economy? In industrial society, the liberal capitalist system has so far proved to be the most efficient socio-economic system. It was in this form that trade unions developed, to overcome the shortcomings of the system and prevent an inadequate distribution rate for labour, and raise the wage level.

The last phase of this process – consumption – means the physical consumption of material goods by individual persons, an entirely non-social act. What I want to stress here is that in industrial society, oriented toward the satisfaction of material needs, the liberal capitalist economic system has proved to be the most efficient social system; the public has given priority to this over other socio-economic and political systems, so long as material productivity grows to develop society and raise the level of material consumption. It can be said that because of this material satisfaction the demand of citizens for fuller participation in politics is reduced to a minimum.

However, in the information society, where the demand for self-fulfilment will become the motivation for action, the process of satisfying the people's demand for attaining objectives will find fulfilment in the production and utilization of information, the selection of action and the attainment of set aims. An economic system such as is most appropriate for promoting, expanding and improving our information productivity, and for promoting, expanding its use, will have to be a synergistic economic system based on synergetic production and joint utilization. This has already been discussed in chapter 7, 'A Synergetic Economic System'. The latter part of this process – selection of action and attainment of objectives – concerns the effects on the external environment of action taken by the doers, and is therefore closely related to the social and political fields.

This is the way in which, in the information society, people's desires will change direction toward the attainment of objectives, which will mean that their demand for participation in decision-making and the management of the economic, social and political system will become stronger.

The second reason is that *the powers of the state and of commercial enterprises have greatly expanded*, and that *policy decisions made by such massive organizations cannot have far-reaching effects on the lives of ordinary people*. For instance, such issues as nuclear power generation, pollution, inflation – every one of these questions bears directly on the lives of ordinary people. Nevertheless, it is only by holding a national or local referendum that citizens can participate in policy decisions on any of these vital issues.

The third reason is that *many of the questions that we have to decide are matters that concern all mankind, global issues that know no national boundaries, and the settlement of which directly affects the lives of all persons*. Take two examples: the question of the population explosion, and shortages of natural resources and energy; these diverse global issues override national borders, and the activities of the United Nations and other international organizations through international cooperation will have an important role to play in resolving them. But by far the more important role will be the voluntary cooperation of all citizens in resolving such global questions, by exercising restraints in their own lives. Take, for example, the question of the population explosion; this is a problem that can only be democratically resolved if throughout the world there are voluntary restraints accepted in the people's way of life, by which they restrict themselves to a basic replacement rate averaging about two children per couple. The only way to get cooperation in adopting such a principle is to secure the participation and agreement of citizens in working for such solutions.

The fourth reason is that *the technical difficulties that until now have made it impossible for large numbers of citizens to participate in policy making have now been solved by the revolution in computer-communications technology*.

One of the major factors that has stood in the way of direct participation of ordinary citizens in national policy making was technical. To consider any such proposition would have involved the work of a great number of personnel over a long period of time, and at tremendous cost. This becomes clear if we recall how a national referendum is held. But now the remarkable development of computer and communications technology has solved this problem at one stroke. The development of communication satellites in particular, and home computers, along with the time sharing systems, together offer a solution to the problems of personnel, time and costs. Furthermore, citizens would be enabled to participate not merely once but repeatedly, enabling them to understand more deeply, from many angles and in a long-range perspective, both the nature and the implications of the problems arising on any issue. From this cooperation would come the fairest, the most reasonable composite solution, so that a final solution from among all proposals will come from understanding and popular consent. We can add that the people would be enabled to participate from time to time in dynamically changing any solution adopted, taking into consideration the actual results of implementation of their selection, and consistent with changes in the objective situation.

Six Basic Principles

To enable this direct, participatory democracy of citizens to function effectively, it will be necessary to set the following six basic principles, which would have to be strictly and faithfully observed.

1 *All citizens would have to participate in decision-making, or at least the maximum number.*

All citizens interested directly or indirectly in any question proposed would have the right to participate in this system, irrespective of race, religion, age, sex or occupation. It will be necessary to ease restrictions on the score of age, with the present voting age substantially lowered, to take in teenagers, depending on the questions to be decided. No democratic solution would be possible without the participation of the teenage generation on such matters as smoking, education and sex for example.

2 *The spirit of synergy and mutual assistance should permeate the whole system.*

To ensure the smooth management of the system of participation, and so that it may be fully effective, the basic attitude of all participating in this system should be inspired by the spirit of synergy and mutual assistance. 'Synergy' means that *each person cooperates and acts from his or her own standpoint in solving common problems* and 'mutual assistance' implies *readiness to voluntarily sacrifice one's own interests for the common good, to level out the disadvantages and sacrifices to other persons and/or groups.*

But synergism goes beyond cooperative effort. Synergistic co-operation brings a wider law into operation, in that the total effect of things acting together is greater than the sum of individual or separate effects achieved.

Parliamentary democracy as we now know it is a system by which the majority imposes its will and policies on the minority, according to the principle of majority rule. It is based on the spirit of individualism and egotism, a self-centred and aggressive attitude that needs to be radically changed to one that is altruistic and cooperative. This is not to be confused with mere collectivism, as it is wholly based on respect for each individual's freedom and interests.

3 *All relevant information should be available to the public.*

When a question is to be resolved with the participation of all the citizens, all relevant information must be made public. In present industrial society, on questions most closely related to the living of the people, the major part of the relevant information is withheld on the grounds of national security or protection of enterprise secrets. But in the future information society, the principle of making information open to the public is a fundamental condition for citizen participatory democracy.

It is necessary that the public be informed not only on factual information, but also on all the possible social, economic and other effects on the lives of the people. Only in this way can each individual understand the problems in which he is interested, not one-sidedly or in the short-term but with a broad, long-term perspective, and participate in decision-making not only from his own standpoint but from the standpoint of the whole community.

In addition, *people will be expected to provide information voluntarily to contribute to a solution of any question.* In the information society, it will be rather this kind of information provided by citizens that will play the major role as basic data for the solution of various problems. Let us imagine that GIUs (global information utility systems) are formed and operated by world citizens. This would mean that information monitored at hundreds of thousands of points on the earth about meteorological conditions, air pollution, plant growth and other matters would be available daily to GIUs. There is no doubt that data so collected would prove to be a fundamental factor in helping mankind meet the changing meteorological and ecological conditions all over the earth.

4 *All benefits received and sacrifices made by citizens should be distributed equitably among them.*

All problems that require participation for their solution are complex by their nature, and the way they are solved would affect different people differently, depending on their place in life and the circumstances in which they live. There would frequently be extreme cases in which those who are to benefit greatly from the solution of a certain problem will be sharply opposed by others who expect it not to profit them. People may also have different evaluations of a problem according to the value systems they cherish. Therefore, in solution of a problem that has a complex effect on the people, consideration must be given to the benefits received and sacrifices made as a result of its solution being distributed in a balanced way among such individual citizens and groups of citizens as are specifically interested in it in one way or another. The balance to be maintained in such cases would be by *combinations of the various benefits and sacrifices*, determined by *the nature and degree of effects* on individuals and groups in different places and positions, a balanced combination achieved with *a long-range perspective*. For instance, sacrifices would have to be balanced by compensations not only in terms of goods or monetary gains, but also, for example, in special educational opportunities for the children of those making the sacrifices; good offices that help in securing desired jobs or positions, or by some other non-economic means. In some cases, long-term political consideration will have to be given so that those who may have to make sacrifices without compensation in solving a certain problem may derive future benefit from the solution of other problems.

5 *A solution should be sought by persuasion and agreement.*

A decision on the solution of a problem should, in principle, be made by the general agreement of all citizens concerned. Patient efforts will be needed to lead opposing individuals or groups to reach agreement. In our present parliamentary democracy, political deals and compromises are made in many cases between the ruling and opposition parties until a decision is reached on important policy matters, or action taken without agreement based on the principle of majority rule. But the information society will require that decisions be made by the *consent* of all participating citizens, and by *persuasion*.

But to make this possible, the decisive point will be the above-mentioned conditions – the spirit of synergy, mutual assistance, publication of all relevant information and a balanced distribution of benefits and sacrifices. In case there are individuals or groups that, even after tireless persuasion, are still opposed to a proposed solution, a second solution put up by such individuals or groups would have to be adopted, out of respect for a minority view, the condition being, however, that such a solution does not impose hardship on other individuals or groups.

6 *Once decided, all citizens would be expected to cooperate in applying the solution.*

The principle is that all citizens will be expected to cooperate in the implementation of a solution decided with their participation. This obligation is a corollary to the right to participate directly in policy making, but it carries with it *the expectation of voluntary self-restraint*, and *should not assume the form of compulsion accompanied by punishment by enforcement of law*, as in the present industrial society. Participation in decision-making, and the observance of the decision through voluntary self-restraints, are inseparable, and it is on this that a new policy-making system and a new social order will be based in the information society. Naturally, there will arise the problem of citizens who violate this social ethical code, but such violators would not be subject to punishment, but would be required to compensate by making *a social contribution*, or by rendering service to the community in some way to make amends for their failure to abide by the decision. The system would not be adjusted by punishment but by the reform of offenders.

Problems Concerning Participatory Democracy

Even if these six principles were to be observed, a number of basic questions would still have to be resolved for participatory democracy to function effectively.

The first question is *how to create and make available accurate and fair information*. More important than anything else for people to make proper policy decisions in a direct, participatory democracy is for accurate and fair information to be made available to them. If an organ to create and provide information were to be monopolized by a small group with a vested interest, and operated to serve that interest, or if specialists and engineers working for such an organ should process information according to their own specific values or ideology, citizens would be participating in policy making on distorted or inaccurate information. To guard against this, the participation of the people in the management of such an organ would be absolutely necessary.

The second problem is *how will the people be able to participate in the settlement of problems that involve state sovereignty?* How can they participate in decisions on questions involving national defence or even war?

There have been recurring controversies over the question of policy making on this kind of question between the military and a president or prime minister, or even between a democratically elected body and the military. The main issue in such controversies is that of the need for instant action in the declaration of war, or in taking defensive measures in the face of enemy attack, or the disadvantage a country faces if a long time is required to make a decision on defensive military action.

But speaking from the basic concept of participatory democracy, it is precisely such important problems of the state that determine the future of a country, and in which all citizens are expected to participate. The only way to resolve this is for *the citizens to participate in peacetime* in making decisions that will prevent a war from breaking out. As citizens come to participate in decisions on measures aimed at prevention of war, such as reduction of military spending and increased aid to developing countries, arguments calling for immediate military action will become increasingly untenable. But for these and other measures for the prevention of war to be really effective, citizens not only in the one country, but also in countries that are in hostile relations with it, will also be required to take concerted action, which stresses the need for international *or global citizen participation.*

The third problem is *how to deal with a problem that cannot be solved even by respect for a minority.* The most important aspect of policy making by participatory democracy is that it aims at many-sided, complex solutions that take into account minority opinions, rather than seeking a single solution as in a parliamentary democracy. This method makes it possible to win the support of a majority of the citizens; but if a handful of citizens should stand out against a proposed solution, how far can the principle of respect for the minority be carried? There is only one way to settle this question. It is to carry out thorough *enlightenment and education such as will lead the citizens to adopt a spirit of synergy and mutual assistance*; the radical way of solving this is by directing such enlightenment and educational work not only at individuals or groups who may be opposed to such and such a solution, but also at all people from childhood onwards, in their homes, in schools and in all their fields of activity.

Together with an objective-oriented pattern of behaviour, this synergism and spirit of mutual assistance among the citizens is the most important for the information society.

9

Computer Privacy:
a Copernican Turn in Privacy

Computer Privacy from the Standpoint of Developmental Stages of Computerization

Human rights in relation to the computer revolution have been discussed mainly from the viewpoint of the invasion of privacy, but the following two points show that this kind of discussion does not meet the case. First, this issue is discussed in terms of the present problem of computer privacy, but the *long-term viewpoint is overlooked*. The second point is that this kind of discussion completely ignores the fact that *information democracy will improve as information productivity increases*. This latter point is the decisive one.

The human right to information, that is, *the human right to know*, along with *the human right to protect secrets*, will be guaranteed by an increase in the production of information, and the human right to protect secrets is an issue involving the very right to privacy. In the information society, the human right to information will inevitably change both in nature and content in the course of the computer revolution.

Viewed thus, let us discuss computer privacy in its relation to the development of the information society from a long-range standpoint, and at the same time examine how this issue can be dealt with, especially in connection with changes in values and social and economic systems, such as can be expected to occur in the course of the development of the information society.[1]

In making this approach, it is necessary to introduce the concept of four developmental stages of computerization: big science-based, management-based, society-based and individual-based, which have already been described in chapter 2.

No Problem in the First Stage

The issue of computer privacy does not arise in the first stage of computerization. In this stage of big science-based computerization, detailed personal data will have been collected on scientists, specialist engineers, military men and civilian staff members engaged in national defence and space development projects. Such data cover birth, personal history, character, specialities, achievements, behaviour, etc. of the persons concerned.

Such personal data records are strictly guarded by the government or military agencies directly under its control, limiting the possibility of such data being stolen or abused by a third party. Information on such persons is directly related to the maintenance of state secrets, and therefore the prevention of any invasion of privacy is the responsibility of the state.

Further, the question of privacy closely related to *national interests existed even before the question of computer privacy arose,* and strictly speaking, was not a problem that should be taken up especially as one of computer privacy. It can therefore be concluded that the question of 'computer privacy' does not arise at the stage of big science-based computerization.

Not so Serious in the Second Stage

It is in this second stage, management-based computerization, that computer privacy in the information society *arises as an important issue.* At this stage, various kinds of personal information are recorded and accumulated in large quantities by computers in the course of computerization under management control by government agencies and private enterprises.

Computers at this stage are used extensively for the automatic processing of large quantities of simple clerical work, repeated on a routine basis. For instance, government agencies introduce computer systems widely and effectively for the quick processing of work, and for the reduction of personnel costs; this includes the collection of taxes and the handling of social security work; and in enterprises, for keeping records of bank deposits, collecting life insurance premiums, electricity charges, etc. Personal data files of this kind previously existed in the form of card indexing and other systems, but the personal data filing system by computers has a number of merits which the conventional card system lacked, such as (1) quick retrieval of specific personal data, (2) integration of personal information from a number of data files, and (3) convenience in copying and transferring such data files.

This opens up the possibility of computerized personal data files *being misused for purposes other than originally intended*, either by the owners of such files or by a third person. This is the main reason for the anxiety about computer privacy.

The characteristics of computer privacy in this stage of management-based computerization can be listed as follows:

1 The purpose of personal data files calls for efficient management.
2 Data on ordinary citizens are filed by the central government and local governments and by enterprises.
3 The filing of such personal data is based on a legal agreement or contract.
4 Data are filed according to predetermined common items.
5 Data filers are, in principle, the users of the data files, and the filers have exclusive use of such files, but those whose data are filed do not directly benefit from the data.

The issue of computer privacy in this stage takes the following three different forms:

1 Government controls over ordinary citizens are strengthened, and the freedom of citizens is violated if the central and local governments use personal data for purposes other than originally intended.
2 Ordinary citizens suffer psychological, social or economic damage if enterprises use managerial personal data and administrative personal data to their own advantage.
3 The right of ordinary citizens to privacy is violated if those personal data are disclosed to a third party.

In this stage, computer privacy has *a quantitatively broad application affecting all citizens*, but, at the same time, *the effect is limited* to the range of personal data recorded. For example, if an enterprise uses the registered names of residents for addressing direct mail, the objections of ordinary citizens do not go beyond the psychological irritation of receiving a lot of direct mail, sometimes called 'junk mail'.

Computer privacy in this stage actually embodies the following two points:

1 If a unified national system of code numbers for all citizens is adopted, the danger will be that the central government or a local government may integrate all kinds of personal administrative records, and utilize such records for the purpose of thought control, or for other political purposes.
2 Even if the harm to ordinary citizens is relatively slight, there is the danger that more serious invasions of privacy will be possible if slight infringements are accepted without protest.

The following two measures are therefore proposed to protect computer privacy at this stage:

1 The government should not adopt a unified national system of code numbers for all citizens without *an effective watchdog organ being set up by the citizens*.
2 The government, a local government or an enterprise, *should not be permitted to utilize or allow a third person to utilize personal data files other than as originally intended*.

If such measures are strictly enforced, computer privacy in this stage can be fully safeguarded.

Decisive Effect in the Third Stage

As computerization enters its third stage, the stage of society-based computerization, the question of computer privacy changes in both character and content. This is because computers at this stage will be used not only in management control by government and enterprises, but also in wide social fields. Computerization will spread into such broad areas as traffic control, pollution prevention, medical care and education, and it will be computer files on personal information relating to these social activities that will have *a direct bearing on computer privacy*.

The following changes in the character and content of the issue of computer privacy will occur:

1 Personal data files will be compiled for the improvement of the social welfare of ordinary citizens and for the advancement of public interests.
2 The filers of personal data will be confined to the government, local governments and other public organs.
3 Personal data will be filed on the basis of a legal obligation or on a contract basis.
4 Personal data will be filed according to approved detailed items concerning the private and social life of individuals.
5 Data files will be utilized by both the data filers and those about whom data is filed.
6 Those whose data is filed will be able to benefit from the files.

The nature of the privacy issue will thus change as follows:

1 If the state should misuse personal welfare data for purposes other than originally intended, government controls on particular persons or groups of persons would be strengthened and *a citizen's right to freedom seriously violated.*
2 If an enterprise should utilize these personal files for management control, *particular persons or groups would suffer considerable psychological, social or economic harm.*
3 If existing general administrative records or enterprise records should be combined with these welfare records, the scope of privacy invasion would be widened and its degree deepened.

Computer privacy in the stage of society-based computerization would be confined to particular persons or groups of persons, and in this respect, the scope of this problem would be narrowed down, but its effect would become incomparably greater than in the stage of management-based computerization. Thus, *particular personal data of one kind can have a decisive effect on the life of persons whose data is filed.* By way of example, detailed records on the history of an illness of a person could deprive that person of an opportunity for employment. Further, school records or behavioural records of an individual from childhood could reveal character, ability, thought and social background, and could be a decisive factor in determining a future career, including employment, marriage, etc.

The following two measures would be necessary to safeguard computer privacy:

1 File management and utilization of personal data on welfare matters should be *under the control of a committee consisting of representatives of public organs and citizens*, with the latter comprising an overwhelming majority of the committee.
2 *A series of social and economic measures would be needed to prevent anyone from suffering social disadvantages* that could arise from personal data files on welfare.

Measures against privacy invasion in this stage could be taken a step further by provision of legal countermeasures, or by a citizens' watchdog organization; it would also be necessary for *the representatives of citizens to participate directly in the management of personal data files*, with the citizens empowered to supervise the files. Further, various social and economic measures would be needed to prevent anyone suffering social disadvantages from the existence of the files. This is of special importance. In more concrete terms, it will be necessary to introduce *a system of social compensation* by which anyone suffering a social disadvantage may secure an opportunity of employment and a stable social life. It would be possible to resolve the problems of computer privacy in this stage by introducing such measures and creating a societal environment in which such measures would be adopted.

Copernican Turn in the Fourth Stage

In this stage, the information utilities will play a leading role in the information society. By 'information utility' I mean a public information infrastructure available to everyone at low cost, at any time or place. Each individual citizen will be able to obtain the needed information, solve problems and seek future possibilities by the creation of highly sophisticated information, merely by connecting the home terminal to the information utility. It is in such a society that all citizens will be able to experience the full blooming of the flower of life. In this stage of high mass knowledge creation, computer privacy will *change again in character and content* as set out below:

1 Personal data files will be used by people to solve the problems of individuals and for the development of their own opportunities.

2 The filers will belong to the public infrastructure.

3 The filing of personal data will be by voluntary registration.

4 Personal data files will cover all items relevant to the private and social life of individuals.

5 Users of personal data files will be restricted to those whose data is filed.

In this case, the preservation of computer privacy will become more serious in character and content in the following ways:

1 If the government should misuse personal data for administrative or political purposes other than originally intended, *all citizens could be completely regimented and deprived of their freedoms.*

2 If an enterprise should utilize such personal data files for the purpose of management control or its own advantage, *all citizens would suffer decisive psychological, social or economic and political damage.*

In this way, if such personal data files should be misused, especially by groups in positions of power, the information society would be turned into a fearful regimented society, and the right of citizens to freedom and protection against privacy invasion would come under a completely polarized, antinomistic relationship.

If personal data files are utilized for the benefit of the persons concerned, they will *bring inestimable benefits* to them; but if abused by other persons or organizations, the files would become a *means of completely controlling the people's destinies.*

The following essential preconditions will be necessary to prevent computer privacy invasion in this stage:

1 Information utilities should be *completely controlled and managed autonomously by citizens.*

2 *New political, economic and social systems should be introduced* to make it possible for such autonomous controls of personal data files to be exercised by the people.

This latter precondition is especially necessary. By new systems I mean such new political systems as will be established in concrete terms by a transition from the present parliamentary democracy of indirect participation to *a democracy of direct participation*, and a new economic system through the transition from our present so-called liberal economic system centring on private enterprises to *a synergetic economy*, in which the infrastructure will play a leading role. Such a new social system will be realized through a transition from the present system which centres on material consumption, to a lifestyle centred on *the creation of knowledge*. In short, the future information society will need to be a citizens' society of a new type, a participatory, synergetic and knowledge-creating society.

Put another way, it will be within the framework of a citizens' society of this new type that information utilities will be autonomously controlled by citizens, and all individual citizens will be free to let their creativity express itself. And if the information society is transformed into a citizens' society of this new type, computer privacy will of necessity undergo a basic change, because, in a synergetic society of knowledge-creation and under conditions of complete, autonomous control of information utilities by the citizens, it will be to the advantage of all citizens to make their own personal data files available in the information utilities as openly as possible, and for all to utilize each other's personal files jointly. This will ultimately make a great contribution to the citizens' society as a whole.

In other words, records on the solution of problems and opportunity developments of individual citizens, and the data gathered for this purpose, will also be of great value to other people for problem solving and for opportunities development. Not only that, these data will be important for the solution of problems that are common to all citizens.

Computer privacy will thus undergo a radical qualitative change. The issue of privacy, which arose as a fundamental human right in the course of the development of modern civilized society, will *lose most of its historical significance*.

Even the human right to information will change drastically in character. *The human right to know will change into a human right to utilize information and the human right to protect secrets will change into a human duty or ethic to share information.* This can be described precisely as a Copernican turn in personal privacy.

Conclusion

Computer privacy changes greatly in nature and content according to the stages of development of the informtion society. Table 9.1 sets out

Table 9.1 Changes in computer privacy, viewed from the standpoint of developmental stages of computerization

	1st stage Big science-based	2nd stage Management-based	3rd stage Society-based	4th stage Individual-based
Forms of personal data files	National security files	Management data files	Social data files	Personal data files
Purposes of files	Maintenance of state secrets	Efficient management control	Increased social welfare	Development of individual opportunities
Filers	Government, the military	Government, local government, enterprises	Government, local governments, public organs	Public organs
Those whose data are filed	Limited number of specific persons	Ordinary citizens	Ordinary citizens, local residents	Local residents
File items	Detailed information on individuals	Formulated common items	Formulated common items, and special items	Non-formulated special items
Beneficiaries of files	Filers	Filers	Filers and those whose data are filed	Those whose data are filed
Privacy invasion Scope	Small (limited number of persons)	Great (ordinary citizens)	Medium (specific persons, groups)	Great (ordinary citizens)
Degree	Infinitesimal (hardly any privacy invasion)	Small (psychological)	Medium (social, economic damage)	Extremely great (regimented society)
Measures against privacy invasion	Direct control by government, military	Legal restraint Technical measures	Joint committee of filers, those whose data are filed	Autonomous control by those whose data are filed
		Watchdog organs	Social measures aimed at reducing the disadvantages of those whose data are filed	Transformation of societal systems aimed at eliminating violations of the privacy of those whose data are filed

these changes in computer privacy, and the following four important tendencies can be read from it:

1 There is a tendency for the purpose of personal data files to *change gradually from national and enterprise interests to social interests*; and in keeping with this, the beneficiaries of data files will also change from the state, the military and local governments to local communities and individuals.

2 The scope of privacy invasion tends to widen from a limited number of persons or groups to ordinary citizens, and *the degree of privacy invasion to increase from a minimum to a maximum*.

3 Further, measures against privacy invasion tend to develop *from legal restraints* on the utilization of personal data files *to citizen participation in and autonomous control of information administration organs*, and *to the transformation of societal systems and institutions*. That is, from technical measures to the control and transformation of social institutions.

In the course of these changes, the nature of the issue of privacy itself gradually changes, and the possibility is that ultimately *it will lose its historical significance*.

Notes

1 Y. Masuda, 'Privacy in the Future Information Society', *Computer Networks*, Special Issue, Amsterdam, North-Holland, 1979.

10

Information Gap: Simultaneous Solution of a Dual Gap

The Information Gap as a Dual Gap

Coupled with the problem of the invasion of privacy by computers, which is certain to be faced in the move toward the information society, there is the problem of an information gap. This information gap exists as an information technology gap between industrialized and developing countries, a gap more serious than the present industrial gap that separates them. The industrial gap is one of productive technology, the main obstacle to its solution being the lack of financial resources in developing countries; but the information gap means the relative absence of information processing and transmission technology in those countries, to which must be added the human factors of levels of intellectual development and behavioural patterns in such countries. These, more than the lack of financial resources, are a major obstacle to technological transfers.

The problem is all the more serious because *the information gap overlaps the industrial gap*, together forming a dual technological gap. If this continues to exist between industrial and developing countries, serious anarchical antagonisms will arise to plague human society.[1]

First, there will be *increased cultural discontinuity*. We are aware of the serious gap between the peoples of underdeveloped countries and the peoples of industrialized nations, but amongst such peoples of underdeveloped countries there is already positive curiosity and interest in the products of the advanced countries, and such people can now board a plane to visit the industrial world if they have funds. But the societal gap of the future between industrialized and developing countries could be incomparably more serious, and mean even complete cultural discontinuity, because the gap is one in intellectual communication essential for mutual understanding.

The second confrontation would be due to *the exhaustion of natural resources*, plus *the environmental disruption* and *the population explosion*. Many developing countries have set out to reach their national goals of industrialization, modernization and improved living standards. But if developing countries push ahead with industrialization by pursuing the course followed by the industrialized countries in the past, it will be all the more difficult to resolve current problems, already becoming very serious, such as the exhaustion of natural resources, the destruction of natural surroundings and the population explosion, which could precipitate humanity into a catastrophic crisis.

In the autumn of 1978 some US scholars engaged in research on nuclear fusion held free discussions on the question: 'What will become of human society if nuclear fusion is developed as a practical technical achievement, and if, in the coming and following centuries, energy can be supplied sufficiently using as material the inexhaustible heavy hydrogen in sea water?' The conclusion reached was as follows:

If humankind is foolish enough to use energy so freely, on the grounds that there is no longer need to worry about it, and if the consumption of energy is increased by 10 percent annually, the total amount of energy consumed on the earth will amount over 100 years to the same as the sun's energy received on the surface of the earth. This would render the planet no longer habitable.[2]

Third, the development of the information society in industrialized countries could be promoted in *an extremely distorted way, assuming a totally military and undemocratic nature*. The sharpening of contradictions between industrialized and developing countries could result in intensified antagonisms and conflicts over security, especially involving the availability of natural resources among industrialized countries, as well as between them and developing countries. The result would be that industrialized countries would tend to utilize their advanced information technology in the use of communication satellites fundamentally for military purposes.

Further, this technology could be used domestically to restrain and control the free private lives and social activities of citizens. Personal privacy could be freely invaded, and the people's autonomous social and political activities could be suppressed, to make the fearful automated Orwellian state a reality.

Simultaneous Promotion of Industrialization and Informationalization

In order to combat this frightening prospect, it is essential to modify and ultimately eliminate the dual gap of industrialization and information, for which the best, in fact, the only alternative is the simultaneous promotion of industrialization and information technology in developing countries. Simultaneous promotion of information and industrial technology will *not cause any conflict* between these two kinds of technology, and given proper means and methods, it will not only greatly promote the industrialization of developing countries but will contribute toward *the creation of a new international order*, involving advanced countries.

By proper means and methods, I mean the following:

1 It is necessary to establish a system of government-led international technical assistance.
2 In introducing industrial technology, the emphasis should be placed on (1) *pollution-free*, (2) *resources conservation* and (3) *intellectual labour-saving technologies*.
3 In introducing information technology, special attention will have to be paid to (1) *the construction of the information infrastructures* and (2) *the introduction of social information systems*, including medical and educational systems.

In regard to item 1, it must be emphasized that no attempt should be made to transplant a uniform system in developing countries, but that *various different forms of technical aid* should be provided, consistent with the wide diversity of the countries concerned.

There is great diversity in developing countries, but they are capable of being classified into two major groups: (1) those countries that have already completed the initial stage of industrialization and could be called half-developed countries and (2) those that are still to begin industrialization. The latter group can be subdivided into (a) countries rich in natural resources and (b) poor countries that depend mostly on agriculture and livestock raising.

It is necessary to take all such conditions into consideration rather than mechanically introducing new technology to them.

In the case of half-developed countries, it may be possible for them to follow *the policy of national capital cooperating with foreign capital*, and for countries rich in natural resources to pursue *the policy of state capital introducing advanced technology*, and for poor developing countries to *receive technological assistance from developed countries*.

In reference to item 2 (countries rich in natural resources), *labour-saving* (engineers and managers) *industrial technologies* are extremely important to developing countries for the establishment of *a new system of international division of labour*, and also for the economic development of the countries concerned.

It is expected that in a new system of international division of labour, iron and steel and oil-refining industries will be developed mainly in countries where the raw materials exist. Importance will therefore be attached to assistance in *pollution-prevention technology* from an ecological standpoint. Further, the conservation of natural resources is a global responsibility to be shared both by industrialized and developing countries. In the introduction of intellectual labour-saving industrial technology, this is expected to have a dual effect, viz., solving the problem of shortages of engineers and managers without causing social distortions of unemployment, and promoting the development of the economies of developing countries. Industrial technology is inseparable from highly sophisticated information technology. In this sense, the induction of the former may be regarded as a form of information technology.

Item 3 (poor countries without natural resources), which deals with *the construction of the first information infrastructure*, shows that it is necessary to form a nationwide information and communication network. This should not consist merely of communication lines for telegrams and telephones, but should be *a digital communication network* permitting the use of computers for transmission and exchange of data. The reason is that we are now in *an era of electro-communication networks* for complex information media, including telephones, TVs, computers and facsimile systems. It is also necessary that such a communication network should form part of a global communication network, making use of communication satellites.

On the introduction of social information systems, it can be pointed out that in developing countries there are strong potential needs for the application of information technology to social areas, including regional development, medical care and education. But the need for information technology in enterprises is not so great, contrasting sharply with the strong need for such technology in advanced countries. There are hardly any large scale, modern industrial enterprises in developing countries, so the management of information technology in developing countries will be oriented from the beginning toward *its utilization for social purposes*. Emphasis will be on the introduction of information technology to serve such areas as *medical care and education*.

There are acute shortages of medical and teaching staff and facilities in developing countries, and the large scale introduction of medical care and education systems in remote areas, making full use of information technology, will greatly increase productivity by medical care and education. Problems which have been considered impossible to solve in the past – the elimination of illiteracy and the eradication of endemic diseases – can be solved for the first time in this way.

Integrated Results of Simultaneous Development of Industrial and Information Technology

If the simultaneous management of industrial and information technology is promoted in developing countries on the above mentioned principles, the following integrated results will be possible (see figure 10.1).[3]

The first result will be *the narrowing of the industrial and information technology gap*. If developing countries succeed in introducing conservation of pollution-free resources and intellectual labour-saving industrial technology, along with information technology centring on education systems for remote areas, industrial and information technology will be able to be developed rapidly. The raised levels of education in particular will make it possible to train modern industrial workers and information engineers, and so lay the foundations for modern industry and for the future development of knowledge-intensive industries.

The second result will be *preparation for an international ecological order*. Since these countries will be oriented toward the development of pollution-free and material-saving industrial technology, such a development in these countries will surely contribute toward the establishment of an international ecological order.

The third result will be *restraint on population explosion*. The introduction of a medical care system for remote areas will

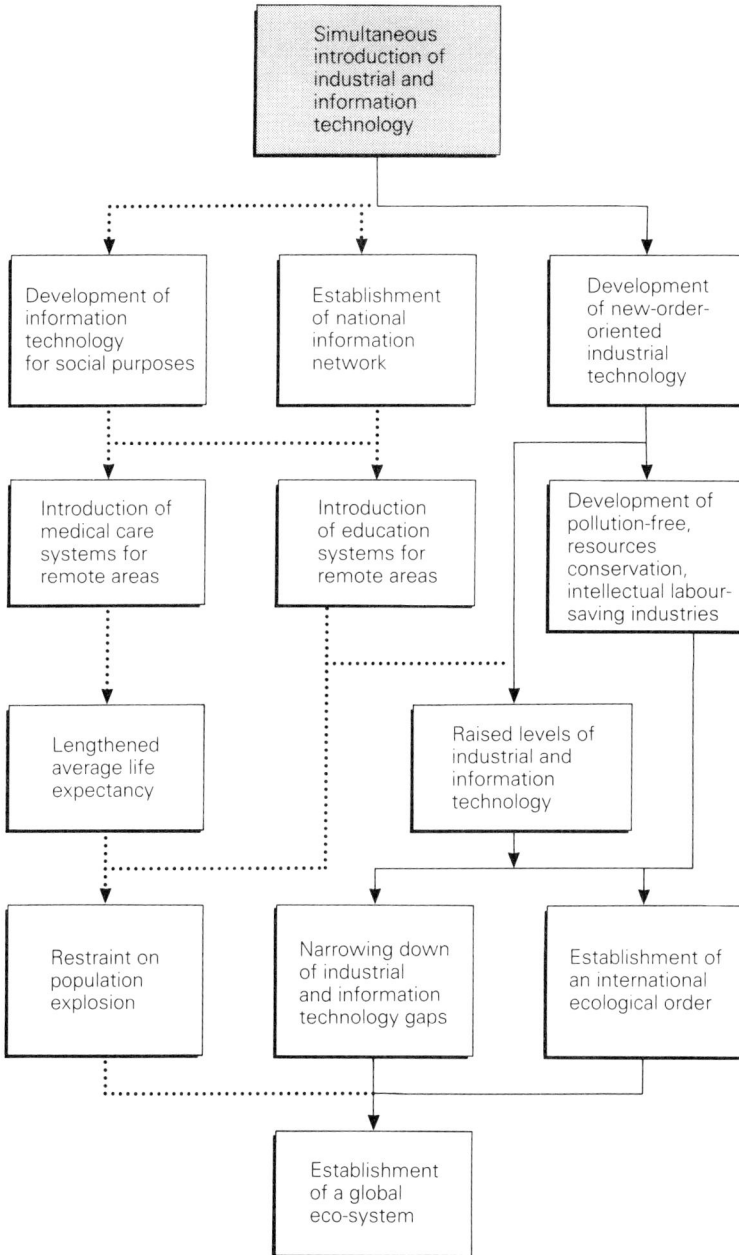

Figure 10.1 Industrial and Information Technology Policy Chart for Developing Countries.

temporarily have a negative effect, in that it will improve the health of the peoples in developing countries and lengthen their average life span. But on the other hand, the raised levels of education will increase understanding of the need for birth control and will make it possible to check the population explosion.

By this means, in the long run, it will be possible to *accomplish the common task of mankind – establish a global eco-system*.

Desirable and Feasible Approach

The approach to a global eco-system is not a mere dream. I mention three points to support the ideal. The first is, as the energy crisis makes clear, the compelling need in the world of today for a new world order and a new system of international cooperation. The second is that pollution-free, resources-saving industrial technology and information technology have been developed, and already many systems utilizing such technological developments are in practical operation. The third point is that information technology is both public and international in character, a point that must receive special emphasis. As has already been repeatedly stated, information is by nature public property; it is non-expendable, non-transferable and has a cumulative effect. Fundamentally, information knows no national boundaries. As communication satellites whirl around the earth, conditions for the formation of a global information network already exist.

It is to be expected that industrialized countries will take the following steps as a desirable and realizable approach:

1 To introduce resources-saving, energy-conserving and pollution-free industrialization, such as will be acceptable to developing countries, and to formulate *a clear-cut action programme* in line with this.

2 For industrialized countries to *reduce their armaments to at least one-half during the 1980s*, and to use the funds thus saved to provide technical aid for developing countries.

3 For advanced countries of the world *to establish an international system of assistance in information technology* of a new type, which could bear the name of IIO (*International Information Organization*).

If industrialized countries and developing countries cooperate and keep in step as they move toward realization of these concepts, mankind not only will be able to avert many crises, such as the exhaustion of natural resources, the population explosion and the like, but the opportunity and perspective of a new global society full of entirely new possibilities will emerge.

Notes

1 Y. Masuda, 'Management of Information Technology for Developing Countries: Adaptation of Japanese Experience to Developing Contries', *Data Exchange*, London, Diebold Europe, April 1974.
2 *Yomiuri Shimbun*, (newspaper), August 17, 1979.
3 Y. Masuada, 'A Plan for the Information Society in Developing Countries', Presentation at the Fifth Brazil Telecommunictions Congress, Sao Paulo, 1979.

11

The Goal Principle:
a New Fundamental Principle of
Human Behaviour

The Material Principle as Traditional

It is necessary that there be a fundamental principle of human behaviour underlying the formation, maintenance and development of human society, which we may call *the societal principle*.

Unlike other animals, man has an innate desire not only for existence but for improvement and better living. Various means of subsistence have been created to satisfy these desires and improve living standards by utilizing these means. This process may be expressed as a process of satisfaction of human needs, as set out below:

> Human needs – production of means to satisfy human desires – satisfaction of human desires.

This cyclical pattern of the satisfaction of human needs moves through a spiral, a qualitative development corresponding to the development of societal productive power.

When a development occurs in societal productive power, the boundaries of satisfaction of human needs expand; new needs arise with the development of the means to satisfy such needs. The basic pattern of the satisfaction of human needs may be plotted as shown in figure 11.1.

So that this cyclical pattern of the satisfaction of human needs may be maintained and developed in an orderly way, it is necessary for human beings as a whole to maintain social existence with some set of common values, and with a common consciousness of the need to

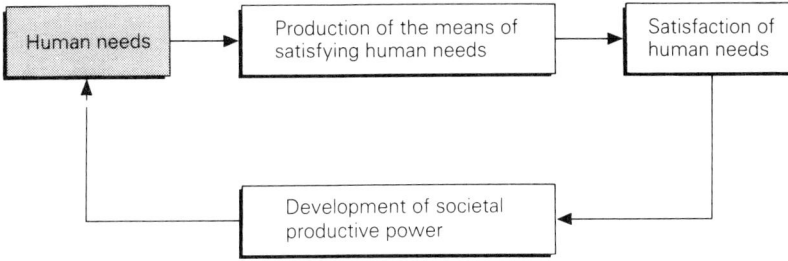

Figure 11.1 Basic Cyclical Pattern of Satisfaction of Human Needs.

realize these values. This basic principle of human behaviour is what we mean by the societal principle.

In the past, the human race went through three types of society – hunting, agricultural and industrial – and the boundaries of human needs have expanded according to the development of societal productive power. Nevertheless, the human race has continued to follow the material principle as the societal principle on which the formation, maintenance and development of human society has depended.

The reason is that human society has maintained a common societal principle bearing the following three features:

1 Human needs have been oriented toward the satisfaction of material needs.
2 The societal productive power, which is basic to the satisfaction of human needs, has been material productive power.
3 The satisfaction of human needs has been achieved mainly through the production and consumption of material things.

Thus, the development of hunting techniques, forming the base of the evolution of hunting society, represented advances in man's capacity to capture animals; which means an increase in material productive power in that it served to increase supplies of food and clothing for each person, and by this improvement in material productive power enabled the urgent needs of life to be met, the need to protect the members of the tribe from hunger and cold.

Then, agricultural productive power, which was the base on which agricultural society was built, represented advances in the capacity to produce material goods in the form of agricultural produce, by which mankind was able to secure food more than enough to sustain life. In these ways, human needs expanded their boundaries from the satisfaction of hunger to the satisfaction of the need for clothing, shelter and the means of everyday living. Still human needs were oriented to the production of material goods.

In industrial society, the enormous development of industrial productive power made it possible to mass-produce goods, greatly extending the boundaries of mankind's material needs, and making it possible for all to enjoy a life of abundance as consumers, rightly called the flowering of material civilization.

Emergence of the Goal Principle

In the information society, however, a wholly different societal principle makes its appearance, viz., the goal principle. This goal principle will have the following main characteristics:

1 Human needs will be oriented toward pursuit of *a self-determined goal*.
2 Societal productive power, which will be the basis for the satisfaction of human needs, will be *information productive power*.
3 Human needs will be satisfied through *the process of production of information and goal-oriented action*.

The traditional material principle and the goal principle may be compared, with respect to their basic cyclical pattern in the satisfaction of human needs, as shown in figure 11.2.

This changeover of the societal principle from the material to the goal-oriented principle will be brought about by *a change from material values to human values*, resulting in part from the shortfall in natural resources, increased pollution and environmental disruption, helped by the remarkable expansion of knowledge-creating information productive capacity, and due in the main to the development of computer-communications technology. We need to note that the goal principle differs fundamentally from the material principle in the satisfaction of needs. There is a change from the pattern of production and consumption of material goods to the production of information, leading to goal-oriented action.

An examination of this basic change in the process of satisfaction of human wants will reveal the structural characteristics of the goal

Material consumption type

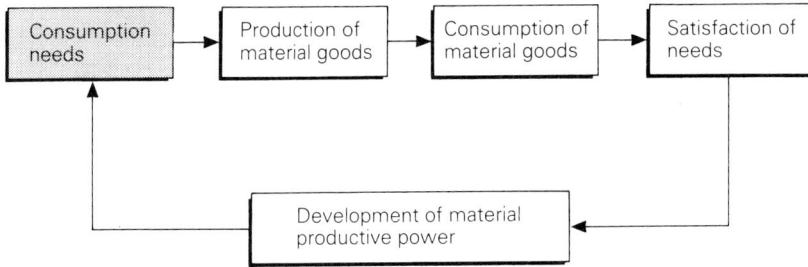

Goal-achievement type

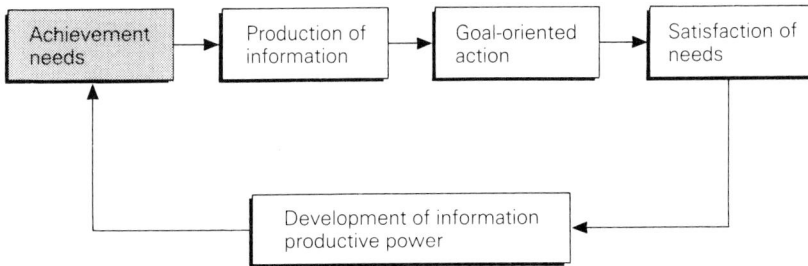

Figure 11.2 Comparison of the Material Consumption Type and Goal-achievement Type Cyclical Pattern of Satisfaction of Human Needs.

principle. These may be examined from two angles, *the logic of situational reform* and *synergistic feedforward.*

The Logic of Situational Reform

What does 'situational reform' mean? I mean the replacement of two processes, the process of production and consumption of material goods, and the process of production of information and goal-oriented action, by situational reform, or the processes of value-realization on the same dimension. Here 'situational reform' means a *process by which the current situation is changed into a new situation that is consistent with the subject's goal*, as expressed by the following formula:

Situation A   → Situation A'   → Realization of a value
(current situation)   (desirable situation) (satisfaction of a want)

Let us apply this 'logic of situational reform' to the process of production and consumption of material goods.

First, the universal concept of economic production is of man making goods that satisfy human wants. That is to say, by the use of physical ability allied to tools and machines, etc., natural resources and raw materials are converted to goods for use. We can define the process of goods production as *the process by which natural resources are converted by goal-oriented action into goods that are useful.*

Let us take the example of hand-made bows and arrows. These are made by converting bamboo etc. and birds' feathers into bows and arrows. By the same principle, consumer durables like television sets are produced. In the case of television sets, natural resources such as iron, copper and oil are first transformed by application of the laws of physics and chemistry into materials which are then made into the many components, and finally these parts are assembled to provide something for use. This reveals a great difference in complexity for goal-oriented labour for the production of goods; but the basis of production is *the change of natural resources into a state in which all goals coincide*, a process which is the same for television as it is for bows and arrows.

What process of situational reform does reaching one's goal-oriented action represent? For example, suppose one sets the goal of 'owning a house to live in' and then goes on to achieve this goal. It will mean a change from the specific situation of living in an apartment to the new situation of living in a house of one's own. From the point of view of goods produced, this is a change into the form of durable goods, i.e., a house. From the point of view of goal-oriented action it can be defined as a goal-conforming change in one's living situation from renting an apartment to house ownership.

Let us take another example. If I set the goal of becoming a department manager in the company where I work, and then attain it, it is a goal-conforming change from one situation to another. Assuming that until then I had been a section chief, to move from the post of section chief to department manager is simply a change in my employment status in the company.

Seen in this way, production-oriented action (economic action for the goal of material production) and goal-oriented action, are exactly the same, in that they are a change from one situation to another, brought about by a subject.

The development, therefore, from the process of production → consumption of material goods to the process of information production → goal-attaining action, may be conceived of as a qualitative development of the *logic of situational reform*. Viewed from this standpoint, *both the material principle and the goal principle can be unified as the situational reform principle*, and the material principle may be encompassed by the goal principle in the broad sense of the word, in as much as the material principle is objective-oriented action for the satisfaction of material needs. The goal principle as used here is therefore, in the narrow meaning of the term, restricted to the satisfaction of goal achievement needs, which are on a plane of higher human desire.

Need for New Interdisciplinary Social Sciences

If this concept of production is adopted, the production of material goods forms only one part of the concept, and broader, objective-oriented acts have to be included.

If a person undertakes an objective-oriented act, a start will be made by gathering the information necessary for the attainment of the set goal or by producing such information for oneself.

Next is the goal-oriented action necessary to purchase and produce goods and tools needed in reaching the goal. Sometimes the goal-oriented action may be sold in the form of labour, the immediate goal being to acquire funds needed to buy the essential goods and tools, and when necessary, part of the goal-oriented action must be used to persuade other people who have something to do with the same field to cooperate in the action. In extreme cases, changes will be necessary in laws, systems and customs, in which case goal-oriented action will have to be given to win the cooperation and support of governments, politicians and citizen groups. Such types of goal-oriented action will necessarily be deeply involved not only in economics but in politics and society as a whole.

When fellow citizens are persuaded to take part in action or to cooperate, the goal-oriented action carried on with others becomes a social action, and the goal-oriented action used to change laws or the policies of national and local governments in the direction desired is fundamentally political action. This concept means that the existing system of sciences, each operating in its own closed domain, will not be adequate as a system of science to meet the needs of the information society in which the principle of a set goal to be attained must operate as the primary principle of action. It will be necessary to establish a new system of interdisciplinary social sciences, in which sociology and behavioural sciences and not economics will hold the predominant place.

Let me emphasize the relevance of today of Max Weber's system of sociology, which set out its basic method as follows:

> We can examine social action, which may appear to be subjective, by objectively examining and regarding motives as being a cause-and-effect relationship.[1]

Weber uses the example of a crowd walking. They may appear to be simply walking, but actually they are going to a horse race. The action of walking may appear to be subjective but is actually objective.

In other words, he considered that human behaviour is motivated and objective-oriented, and it is possible to trace objective cause–effect relationships in the social process if the relationship of means and objective is translated into a cause and effect relationship.

If social action is not impulsive or irrational but is objective-oriented and rational, it is possible to *replace the term objective-means oriented action with the term cause-effect relation*, as Weber says, and as has been said already. The goal principle will be firmly established as the first principle of human behaviour in the information society. Then the social actions of citizens in general will become goal-means relationships that can be regarded as cause–effect relationships.

Four Types of Feedforward System

As for synergistic feedforward, we must begin by examining the relevance of the logic of situational reform to the feedforward system. As already explained, 'situational reform' means a change from the existing situation to a more desirable one, a process in which the subject of action works on the external environment to make it more desirable; this may be expressed as *a feedforward from the current situation to a desirable situation*. By 'feedforward' is meant *control in moving toward a goal*, and viewed from the standpoint of the subject of action, it means *a controlled development of the current situation to change it to a more desirable situation*. And included in this process of feedforward is not only the external environment on which the subject of action works, but also the subject of action itself.

The action for situational reform, viewed from this standpoint, is precisely the process by which the subject of action controls itself to adapt to a desirable situation. In this sense, situational reform is *a simultaneous feedforward process* by the subject of action and its external environment for realizing a desirable situation.

There are various feedforward systems as processes of situational reform. We may classify these into the following four types (see figure 11.3):

1 dependent feedforward
2 controlled feedforward
3 balanced feedforward
4 synergistic feedforward

By 'dependent feedforward' we mean *one by which the subject of action exercises control in attaining the objective*, especially when the subject of action is dependent greatly on the external environment. This feedforward process is a negative and passive one.

The 'controlled feedforward' means that *the subject of action is in control of the external environment in a mutual relationship of forces*, a

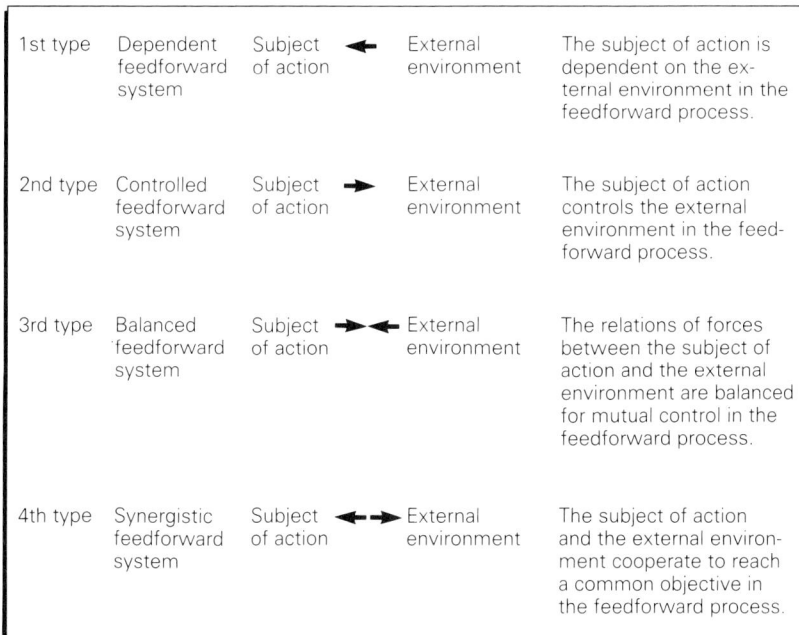

1st type	Dependent feedforward system	Subject of action ← External environment	The subject of action is dependent on the external environment in the feedforward process.
2nd type	Controlled feedforward system	Subject of action → External environment	The subject of action controls the external environment in the feedforward process.
3rd type	Balanced feedforward system	Subject of action →← External environment	The relations of forces between the subject of action and the external environment are balanced for mutual control in the feedforward process.
4th type	Synergistic feedforward system	Subject of action ←→ External environment	The subject of action and the external environment cooperate to reach a common objective in the feedforward process.

Figure 11.3 Four Types of Feedforward Systems.46

process in which the subject of action exercises control to bring the external environment into line with what is desired. This process is active and positive and takes the character of external control.

In 'balanced feedforward' *the subject of action and the external environment exercise mutual control if their relations are in balance.* This could mean mutual control of a competitive or antagonistic nature.

The 'synergistic feedforward' means that *the subject of action and the external environment are in mutually complementary relations and act together to reach a common objective.* This is a synergistic, mutually complementing and integrated process of control.

In human society, however, the feedforward system assumes a far more complicated form, because it involves a feedforward system between the human and the natural environment, as well as between person and person within human society. These are not independent of each other; their relationship is such that the former is primary and the latter secondary and subordinate. In other words, at the stage of human development in which a certain kind of feedforward system between mankind and nature begins to function, a corresponding feedforward system between person and person within human society arises. This is because the development of human society requires a feedforward process between humans and nature that is favourable to humans; and further, in this development of human society, the process of social stratification of human society proceeds gradually.

Tracing the historical development of feedforward systems in human society, one finds an interesting phenomenon; that in human society, the feedforward system began as a dependent one, and is due now to be replaced by a higher synergistic one.

Dependent Feedforward and Taboos

The feedforward system, operating as a situational reform process, functioned in primitive society in the form of taboos. To primitive man the natural environment was chaotic for a very long period of time, but gradually nature and the universe came to be regarded as ruled by a powerful god. From this, the idea that man should obey this god and adapt his own life to a natural order became prevalent in society, and with this came taboos as the first form of feedforward system. Taboos moulded a social order that conformed to the natural order, and in effect were a dependent feedforward principle based on religious norms.

Taboos formed a system of passive dependent feedforward among human beings, on *the concept of divine laws*, leaving no room for voluntary action selection by human beings.

Controlled Feedforward and Ruling Power

In agricultural society a new type of feedforward system operated. This was controlled feedforward in the form of *absolute rule* by one class over another, usually that of a feudal lord over peasants.

Here we see the operation of a feedforward system within human society, the background of which is a partial change of the feedforward system between man and the natural environment from a dependent system to a controlled system. Even in agricultural society, the universe was conceived of as ruled by the will of a god or gods, and the divine rules largely determined human behaviour. So, the dependent feedforward system was deeply rooted in human society in the relations between mankind and the natural environment. However, mankind developed agricultural techniques and succeeded in turning vast expanses of wild land into arable land.

Only the feudal lord and a small class of knights took goal-oriented social action. All the peasants could do was follow these actions and yield to the power of their rulers; they could take no other social action. They were bound to the land and not permitted to leave it. Heredity determined one's occupation. And it was the peasant's mandatory duty to deliver tribute to the lord in the form of a part of the harvest produced each year. The rule of the feudal lord functioned as a single and absolute controlled feedforward law that encompassed all political, economic and social aspects of living.

Balanced Feedforward and Price Mechanism

After this historical development, balanced feedforward originated as a new feedforward system in industrial society. This came into being on *market principles and the price mechanisms* that automatically brought some kind of economic order into industrial society, such as, in Adam Smith's terms, the commercial transactions guided by self-interest in the give and take relations of *Homo economus*, and the changes in market prices brought about by that *Invisible Hand*. At the micro level, there has certainly been balanced feedforward in the sense that the seller and the buyer have maintained economic order by each exercising self-control in economic operations, while bargaining in the market place. And at the macro level there has been balanced feedforward of the macro economy, consisting of the accumulation of balanced micro level feedforward (in the sense that a large number of enterprises are guided by the invisible hand of changes in market prices): supply and demand have balanced and in general, order has prevailed in the overall economy.

Behind this balanced feedforward in industrial society there has been controlled feedforward involving nature, *the control systems by which man has changed natural resources into useful goods.*

Mankind has had the temerity to disregard the supreme life-force in relations with nature, substituting natural science for the rules of nature, and treating the natural environment as an inexhaustible reservoir of resources, and plundering resources in order to achieve a maximum material satisfaction, thus causing widespread environmental disruption and upsetting the balance between man and nature in the ecological sense.

Synergistic Feedforward and Globalism

In the information society, a new type of feedforward – synergistic – will come into being.

To support this synergistic feedforward in the informtion society the first factor will be globalism. The spirit of globalism will establish the importance of the symbiosis of man and nature, and with the change in the spirit of the times will come a conception of ecological systems which harmonizes the systems of nature and of humans to maintain and develop the existence of humanity.

The second factor is that the structure of the information productive power of computer-communications technology has always had the characteristic of synergistic feedforward. Information does not disappear with use, and no matter how many times it is used its existence continues unchanged. Another point is that the structure of production is characterized by man–machine self–multiplication of information. Public usage and the synergistic feedforward of information both result from these two characteristics of the utility of computer-communications information; and further, the utility of information will become more and more oriented toward the public interest because of the great number of people who are able to use it.

In the information society of the future, *the right of usage* is predominant, not the right of ownership, and the principle of synergy rather than the principle of competition will be a basic societal principle.

This explains how the new, synergistic type of feedfoward will come into being in the information society. In the information society of the future individuals will have a common social goal, and as individuals and groups, will build an order of social action among themselves in order to attain their goal by working together synergistically.

The first basic characteristic of this synergistic feedforward will be *the common goal*, based on common awareness and common group needs that do not conflict with the goals of individuals. The second is that action will be *voluntary*. There must be no coercion. The third is that individuals and groups will *cooperate actively* to attain their common goal, cooperation that will be dynamic, not static, in methods and organization. The fourth characteristic is *self-control*. Individuals and groups will voluntarily and continuously control their own actions as they move toward the common goal. These are the basic principles of synergistic feedforward.

In the information society, however, synergistic feedforward will function as a common system for all mankind in relations with nature, and in relations between humans within society. This is the concept of globalism and the spirit of synergism between the divine and the human that unifies two feedforward systems.

Notes

1 H. Otsuka, *Method of Social Science*, Tokyo, Iwanami Publishing, 1966.

12

Voluntary Communities: the Core of the Social Structure

Futurization: the Fundamental Pattern of Life

Let us pass now to the question of the social structure and system in the information society. I believe the core of the social structure of the information society will be *voluntary communities*. Voluntary communities will be *the form of society in which people, of their own choice, will participate in building a community by their own efforts*. The basic precondition for bringing about such voluntary communities is what I have called *futurization*. The word 'futurization' implies *actualizing the future, bringing it into reality*. Expressed metaphorically, it means *to paint a design on the invisible canvas of the future, and then to actualize the design*.

When actualizing one's own future becomes the basic pattern of life in the information society it will be an epochal stage in the development of human society. Qualitative change and development in human society has, until now, followed the development of societal productive power, which means material productive power, the life patterns of people having been oriented toward the expansion of material consumption. This is true of all former types of society: hunting, agricultural and industrial.

In the information society, however, societal productive power will mean information productive power. The expansion and spread of this information productive power will be the stimulus to seek fulfilment of time needs rather than material needs, encouraging the adoption of time-values, and a new mode of social action oriented toward the realization of future time-value.

Individual Futurization as the Starting Point

Let us start with futurization by people as individuals – *the individual futurization stage* – inspiring the initiation of voluntary communities. To realize the time-value of one's own future time, each person sets the target, on the invisible canvas of a desirable and feasible vision, and sets about the task of actualizing it. Having selected the most appropriate action, one then goes on to means-oriented action. This is the individual futurization stage.

This individual futurization implies, however, two self-contradictions: there is *the inevitable conflict with the futurization of others*, as well as the essentially limited scale of individual futurization. Individual futurization, though termed individual, will have to be carried out within the structure of society, and conflicts will always arise between the futurization of one individual and another.

Let us suppose one takes action toward a goal that conflicts with that of another, reducing thereby the probability of an individual attaining the set goal. If, in order to increase the probability of achievement, one tries to actualize futurization by relying wholly on one's own chosen means-oriented action, then *the scope and scale of futurization is so much less*. Such futurization becomes merely a Robinson Crusoe-like self-sufficient futurization.

Another contradiction in the transition from an industrial to an information society will be *the contradiction between employed labour and voluntary labour*. If, in order to have more free time for oneself a person decreases working hours to a minimum, then income from employment will decline as free time increases, and to the extent that income decreases, there will be a decrease of purchasing power for the essential means of futurization.

Reaching the Stage of Group Futurization

Because of the need to break free of the contradictions of individual futurization, *group futurization* will gradually develop. By group futurization we do not mean futurization as a group that overrides the individual, but group futurization that begins only with individual futurization. This form of group futurization expands and develops individual futurization, a stage that can only come about when three conditions are met. First, group futurization must be in accord with the needs and direction of individual futurization, if not directly, then, at least, indirectly; second, participation in group futurization must be voluntary, with no restrictions whatever placed on participation; third, individual social labour in group futurization must always be of a voluntary not controlled nature.

When these three conditions are met, participation in group futurization will not conflict with the needs of one's own futurization, rather one's own futurization design will be actualized within a much wider involvement in society.

Let us consider some examples of this type of group futurization. One such might be venture businesses with *a new type of management organization*. A number of people who share the same business goal might, for example, get together and set up a business venture. With shared risk, each would become responsible in the activities of the enterprise for one's own particular field. One might undertake the responsibility of providing capital; another might contribute specialized knowledge and expertise. Each would have equal rights as participants in management; income from the enterprise would be distributed according to the agreement between all members, and each would agree to share the responsibility in case of loss. The activity of such an enterprise would not result in a return in profits to those providing the capital, but rather, would result in the participation of all members in deciding how to use and distribute the net income. In this type of venture all positions would be decided by the agreement of all participants which means that even the chosen president would be subject to the authority of the whole group, and failure to improve the performance of the enterprise would mean dismissal from such office. And if, for example, it was found that working for the enterprise conflicted with one's individual futurization, leaving the enterprise freely would be the only option. To carry out the responsibilities undertaken for a fixed period with the agreement of all would be the prime duty of such a person, who would not be able to quit before the term of office ended.

Another example could be group futurization in the public arena. Suppose a person wants to actualize the desire to enjoy cycling, towards which end action begins on making a public cycling path in the vicinity. Such action begins from one's own individual needs, but suppose there are other enthusiastic cycling fans in the locality, eager to have a public cycling course. The originator, together with the others, may bear part of the burden of cost for the purchase of necessary materials and tools, and may give free time in constructing the cycling course.

Voluntary Communities Appear

Futurization begins with individual futurization, advances to group futurization, and when it develops to a certain stage, it goes on to develop into *the community futurization* stage, in the formation of voluntary communities.

Community futurization differs fundamentally from the earlier stage of group futurization, not in an increase in the number of people participating or in the complexity of its function, but in that it has now taken on the character of a community. I would call a small social unit composed of a group that comes together through community futurization, a voluntary community.

The 'classical concept of community' that has come down to us refers to *a group of people who carry on life together with a common social solidarity*. Communities have had a long traditional history, the general form of which up till now has been that of the local community. Their fundamental characteristics have been (1) a homogeneous creed for day-to-day living; (2) they are indigenous; (3) they are isolated; (4) there is communal ownership of the means of sustaining life; and (5) there is group life. There has been a strong tendency for communities to share a common religion, to be other-worldly and self-sufficient.

The voluntary communities that we are considering here have several basic features in common with traditional cooperative communities.

The first is that such communities will come about by *voluntary association and be self-made*. People will participate of their own free will and build such communities by their own efforts and labour.

The second feature will be *voluntary management*. This means that the people themselves will become the subject that maintains social order within the group to which they belong; and further, that, on principle, social order will be maintained by the self-discipline of the individuals and the group. Voluntary management here means that the community will not operate on a system of bureaucratic control.

The third feature will be a *sense of mission*. People will share the goals of the community, be bound strongly to each other by a sense of mission, and give mutual support in the actualization of future goals.

The fourth will be *synergism*. People who belong to the community will work together in a mutually complementary way to achieve a shared goal.

On the other hand, voluntary communities in the information society will have several novel characteristics that mark them as differing vitally from traditional cooperative communities.

The first such characteristic is that the voluntary community will become *the most important social organizational unit* of the information society.

There are some examples of communities that have been voluntarily planned and created. The Puritans fled from England to a new continent with visions of creating an ideal society; there was Robert Owen's agricultural community; in Japan before World War II there was Saneatsu Mushakoji's Atarashiimura (New Village); more recently, there are communes in the United States set up by what were called 'hippies', and others. Few of these voluntary utopian communities developed or were able to survive for long.

But voluntary communities of the new age will not be simply unattainable utopias. They will have ample realistic possibilities of success. Citizens will have abundant free time and a high standard of living, and a high level of education. Their way of life will expand from material consumption to futurization, and the socio-economic environment will become very conducive to the fulfilment of their desires. It is at this stage that a synergetic economic system and participatory democracy will become a reality.

In this development, voluntary communities will become widespread rather than being limited to restricted areas, scattered and isolated; they will become the most important organized social units constituting the information society.

The second characteristic of voluntary communities will be their concept of *information space*. In such communities, ties to a locality will not necessarily be a fundamental property of the community as they have been in the past, because the emphasis and binding force will be shared common ideas and goals. Such a community, not being tied to one locality, is a community that has information space, *invisible but perceptible space functionally bound together by information networks* based on computer-communications technology. People belonging to such a community will work together in a mutually complementary way to achieve a shared goal. This emphasizes the strong sense of mission that binds them together in seeking the goals of a community.

In referring to a 'community of destiny' I present the idea of globalism, the concept of this spaceship earth, on which alone mankind can exist, and of a *common awareness* of the succession of crises confronting humanity: the population explosion, shortages of resources, the possibility of nuclear war, racism, the Third World, and widespread poverty and hunger.

Even the roles of individuals will not be subdivided and separated into divisions of labour, as of now, but will take shape as *functional cooperative labour*, with roles changing dynamically in response to the changes in the environmental conditions surrounding the community. The people who make up such communities will need to be able to change their roles in response to current demands, and the assignment of new tasks within the community as a whole. Their functionally cooperative labour which dynamically changes will become the basis of societal development of such voluntary communities.

The third characteristic will be *the growth of multi-centred, multi-layered communities*, not as closed self-contained cooperative communities, but as *open* communities. Each community, while maintaining its independence, will be interlinked to complement others. More than that; groups and individuals making up each community will at the same time be participants in other communities, so that even moving to another community will be possible. It would be impossible for participants always to base their whole lives on one voluntary community, and freedom to seek income from other community sources or from general employment would be necessary. In other words, being a member of a functional voluntary community would not prevent a person from belonging to several communities or engaging in some other paying work.

So, generally speaking, citizens who are members of voluntary communities will belong as well to other social groups, thus being 'multi-belonging' citizens. This is what I mean by multi-polar, multi-layered communities.

Seen in the macro sense, multi-layered voluntary communities will form an important societal sector operating alongside existing business and governmental sectors, and open to the two other sectors, maintaining mutually dependent relations with them, actively

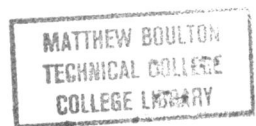

MATTHEW BOULTON
TECHNICAL COLLEGE
COLLEGE LIBRARY

influencing them and ultimately bringing them into homogeneous relations.

The business sector, by establishing *mutually dependent relations with the voluntary sector*, will cease to be a self-sufficient and closed traditional sector, and will become the subject of vigorous development. In the transition from the industrial society to the information society, a high level of material productive capacity of the business sector will have to be maintained, but a fundamental change will take place in the nature, functions and role of the business sector, all basic industries taking on the form of public infrastructures. Further, *participation by the citizens in the management of enterprises* and *autonomous worker control* will become the pattern of management, and the managers will be trained experts in enterprise management.

The governmental sector will be greatly *simplified*, exercising a minimum of powers compared with the present aggrandized bureaucratic governmental organizations, all important policies of the state being decided by direct participation of the citizens. The dominant form of maintenance of public order, replacing resort to the law, will be *the autonomous restraints exercised by the citizens themselves*.

Such changes in the business and government sectors are the outcome of the homogeneity developed with the voluntary sector. When these sectors have become homogeneous with the voluntary sector, and the voluntary sector has become the leading societal form, human society will have been converted into a voluntary civil society, which is what we mean by *the high mass futurization society*.

Voluntary communities can be divided broadly into two types: *the local voluntary community*, and *the informational voluntary community*.

The Local Voluntary Community

This type of voluntary community will come into being in a specific locality, a new development along the lines of existing types of local communities. A prototype of these local voluntary communities is already evidenced by the growth of citizens' movements.

If we regard consumer movements for *self-protection* against defective products as the first stage of citizens' movements, the second stage may be seen as citizens' movements centring on *anti-pollution campaigns* and *litigation*; and the third stage may be seen as *citizen participation in local government* by voluntary citizens' movements. These citizens' movements of the third stage can be seen as nascent prototypes of the local voluntary community. Quantitative changes take place in citizens' movements belonging to the third category, a development from previous citizens' movements, from protectionist activities that carry on campaigns, demonstrations, strikes etc., to independent movements organized by the citizens themselves. At this third stage, active citizens' movements, by *the voluntary efforts of the citizens* themselves will set out to move their own civic life in the direction they desire. The autonomous productive activities that the residents in a given locality carry on for themselves – such as farm production and cultivation by people living in public housing, managing for themselves the roads, parks etc. in the area where they live, the delivery of newspapers and mail by young people and students, and even the disposal treatment of garbage – all such activities will develop into full scale town planning and the construction of a new community by the residents themselves.

The ultimate stage will involve voluntary management of the community. In this fourth stage the management of the local community will be by *the voluntary services of the citizens themselves*. The administration of law and justice over the nation and the state will be *greatly simplified*, and a complete division of powers, with reduced functions of the central administration, will result. It will mean that the substance of local community operations will be concentrated into *smaller societies, each with its own characteristics and individuality*.

There will be communities of the elderly, communities of scientists, and idealistic and religious communities bound together by a common ideal or religion. Probably many other kinds of diverse local communities will evolve, such as communities of sunlovers or nature-lovers.

The Informational Voluntary Community

The fundamental characteristic of such voluntary communities will be their freedom from ties to a local place, a completely new type of voluntary community. The fundamental bond to bring and bind people together will be *their common philosophy* and goals in day-to-day life; it is the technological base of *computer-communications networks* that will make this possible. Communities formed in this way can be described as functional information space connected by information networks.

The precursor to these voluntary communities will be *the voluntary association*,[1] a social group voluntarily formed by people who have similar hobbies, beliefs, or ideals. There are many examples of such voluntary associations in current industrial society. One example is the Club of Rome, international in scope and policy-oriented. To be *international and policy-oriented* is one of the fundamental characteristics of the voluntary communities that will emerge in the information society, because the very framework of the information society is globalism, the prevailing thought of our times: global information networks, and *a goal-oriented knowledge-creating citizenship.*

The Club of Rome, as a voluntary global knowledge-creating group, has wielded a strong influence on all people throughout the world, but it has not yet escaped from the restrictions of a voluntary association. Many more conditions must evolve before this kind of association can develop into a voluntary community. Global information networks have only just come into being, and these are used for stage goals by the larger countries of Europe, the United States and the Soviet Union, and civilian use has not progressed beyond the large multinational corporations, nor have the citizens of the current mass society developed such attitudes as will be necessary for the formation of an informational community; and further, this kind of international, policy-oriented informational community can only come into being when many voluntary communities (small social groups) have already been formed and have linked up in various ways.

As an example of this sort of global, informational, voluntary community of the future, we can imagine what we now call *a zero population informational community.*

Population control will be an essential means of saving mankind from the endemic hunger and poverty we see in some places now, and which could become global by a worldwide population explosion. It is a situation calling for an informational community organized throughout the world to set this problem and goal before all people. The programme of action would be, for example, *no family should consist of more than two children*, and this ideal would be pursued by the families of those participating in such a community. It is important to note that this form of population control will not be enforced by laws and the authority of the state, but by *emulation and voluntary self-control exercised by families and individuals.* To exercise birth control, the member groups would provide publicity, counselling, medical care and the best hospital equipment. The necessary funds to carry this out would be provided by *the contributions of individual members*, by business circles and by governmental aid where necessary.

Some additional points need to be mentioned here. Those who voluntarily participate in the zero population movement will naturally observe the standards set on ideal family size. Then, too, part of one's ability, income and time will be devoted to the movement, with activities of individual participants carried out voluntarily, and independent of government policy, laws, funds and facilities. The accumulative result of this kind of communal action will be a decisive factor in curbing any population explosion, one of mankind's major problems.

My own firm belief is that many such global informational voluntary communities will take shape as we approach the twenty-first century. Among such, the most needed and feasible communities would be 'non-smoking voluntary communities',[2] 'global anti-nuclear weapon, complete disarmament communities', 'global energy conservation and anti-pollution, nature preservation communities'. Globalism, time-value and a goal-oriented mode of action will be the universal concepts shared by these global voluntary communities.

While individually pursuing their own futurization needs through goal-oriented action, people will participate and work together in one or more voluntary communities, and as members of a global community. They will cooperate in solving the problems and crises that are common to all mankind. This is how I see the future information society ultimately functioning.

Notes

1 W. M. Kitzmiller and R. Ottinger, *Citizen Action: Vital Force for Change*, Washington, DC, Center for a Voluntary Society, 1971.
2 *Aktuellt*, Bulletin of VISIR, Stockholm, 'The Smoking Digest: Progress Report on a Nation Kicking the Habit', Washington, DC, US Department of Health, Education, and Welfare, 1977.

13

Computopia:
Rebirth of Theological Synergism

A Vision of Computopia

As I come towards the end of this book about the information society, I want to present you with a vision of 'Computopia' (abbreviation of computer utopia).[1] Looking back over the history of human society, we see that as the traditional society of the Middle Ages was drawing to a close, the curtain was rising on the new industrial society. Thomas More, Robert Owen, Saint-Simon, Adam Smith and other prophets arose with a variety of visions portraying the emerging society. Of special interest to me is Adam Smith's vision of *a universal opulent society*,[2] which he sets out in *The Wealth of Nations*. Smith's universal affluent society conceives the condition of plenty for the people, economic conditions that should free the people from dependence and subordination, and enable them to exercise true independence of spirit in autonomous actions.

Smith presented *The Wealth of Nations* to the world in 1776. James Watt's first steam engine was completed the year before, but although the industrial revolution was under way, Smith's grand vision of a universal society of plenty was still far off when he died in 1790. His vision seems to be half-realized two centuries later, as society reaches Rostow's High Mass Consumption stage,[3] which means that the material side of Smith's vision of people having wealth in plenty is partially accomplished, at least in the advanced countries. His wider vision of the individual independence and autonomy that would follow has clearly not been realized, because the axis around which the mass production and consumption of industrial goods turns in industrial society comprises machines and power. Capital investments are necessarily immense, with the result that the concentration of capital and corresponding centralized power are the dominating factors. This is the fundamental structure of all industrial societies, something that transcends the question of a society being capitalist or socialistic.

Industrial societies are characterized by centralized government supported by a massive military and administrative bureaucracy, and in capitalist states supra-national enterprises have been added that make the modern state dependent on the trinity of industry, the military and the government bureaucracy. In industrial societies the individual has freedom to take social action in three ways. A person is able to participate indirectly in government policy by voting in elections once every few years. One has the freedom of using income (received as compensation for subsistence labour) to purchase food and other articles necessary to sustain life, which implies freedom to use free time on weekends and holidays as one likes. This freedom of selection, however, is freedom only in a limited sense, quite removed from the voluntary action selection of Adam Smith's vision.

As the twenty-first century approaches, however, the possibilities of a universally opulent society being realized have appeared in the sense that Smith envisioned it, and the information society (futurization society) that will emerge from the computer-communications revolution will be a society that actually moves toward a universal society of plenty.

The most important point I would make is that the information society will *function around the axis of information values rather than material values*, cognitive and action-selective information. In addition, the information utility, the core organization for the production of information, will have the fundamental character of an infrastructure, and knowledge capital will predominate over material capital in the structure of the economy.

Thus, if industrial society is *a society in which people have affluent material consumption*, the information society will be *a society in which the cognitive creativity of individuals flourishes throughout society*. And if the highest stage of industrial society is the high mass consumption society, then the highest stage of the information society will be *the global futurization society*, a vision that greatly expands and develops Smith's vision of a universal opulent society; this is what I mean by 'Computopia'. This global futurization society will be a society in which everyone pursues the possibilities of one's own future, actualizing one's own self-futurization needs by acting in a goal-oriented way. It will be global in as much as *multi-centred voluntary communities of citizens participating voluntarily in shared goals and ideas will flourish simultaneously throughout the world.*

Computopia is a wholly new long-term vision for the twenty-first century, bearing within it the following sevenfold concepts:

Pursuit and Realization of Time-value

My first vision of Computopia is that it will be *a society in which each individual pursues and realizes time-value.* In Japan, the advanced welfare society is often talked about, and people are now calling for a shift of emphasis from rapid economic growth to stable growth, stressing social welfare and human worth, sometimes expressed as a shift from a GNP society to a GNW society, i.e. gross national welfare. The current idea of an advanced welfare society, however, tends to place the emphasis on the importance of living in a green environment where the sun shines. Obviously, in seeking to escape from the pollution and congestion of cities, and from the threat of a controlled society, this concept is significant, as indicative of our times. Yet it does not embrace a dynamic vision of the future, which I feel is its greatest weakness. The disappearance of pollution and congestion, or even escape from the cities, will not alone bring satisfaction. Human needs are of a very high dimension that must be actively satisfied – the need for self-realization. The futurization society, as I see it, will be a society in which each individual is able to pursue and satisfy the need for self-fulfilment.

The self-realization I refer to is nothing less than the need to realize time-value, and time-value, of course, involves painting one's own design on the invisible canvas of one's future, and then setting out to create it. Such self-fulfilment will not be limited merely to individuals all pursuing their own self-realization aims, but will expand to include mini-groups, local societies and functional communities.

Freedom of Decision and Equality of Opportunity

My second vision is *freedom of decision and equality of opportunity.* The concepts of freedom and equality grew out of the Puritan Revolution, which reached its peak in England between 1649 and 1660. Initially the ideas of *freedom from absolute authority* and *legal equality* underlay these concepts, backed by the theories of social contract and individual consent as the basis of political authority, theories that maintain that freedom and equality are natural rights for all people. These two ideas provided the theoretical base for the formation of modern civil society.

As the capitalist economic system came into being, freedom and equality developed conceptually to include 'freedom to work at something of one's own choice', 'equality of ownership' and 'freedom to select an occupation', and 'industrial equality', more commonly referred to as free competition.

The information society will offer new concepts of freedom and equality, embodying *freedom of decision* and *equality of opportunity*.

As I have said, the information society will be one in which each individual pursues and realizes time-value. In this type of society the freedom that an individual will want most will be *freedom to determine voluntarily the direction of time-value realization in the use of available future time*. Call it 'freedom of decision'. Freedom of decision is the freedom of decision-making for selection of goal-oriented action, and refers to the right of each individual to determine voluntarily how to use future time in achieving a goal. This will be the most fundamental human right in the future information society.

'Equality of opportunity' is *the right that all individuals must have, meaning that the conditions and opportunities for achieving the goals they have set for themselves must be available to them*. This will guarantee that all individuals have complete equality in all opportunities for education, and the opportunity to utilize such opportunities for action selection. Guaranteed equality of opportunity will, for the first time, assure that the people will share equally the maximum opportunities for realizing time-value.

Flourishing Diverse Voluntary Communities

My third vision is that there will be a *flourishing of diverse voluntary communities*. A society composed of highly educated people with a strong sense of community has long been a dream of mankind, and several attempts have been made to bring it into being. Recently communes have been formed by groups of young people, and a

number of cooperative communities have been formed in Japan. One
was the *Yamagishi-kai*, formed after the war. The rapid growth of
information productive power built around the computer will see
some big advances and developments beyond the ideas and attempts
of the past. There will be enhanced independence of the individual,
made possible for the first time by the high level of the information
productive power of the information society. The development of
information productive power will liberate man by reducing dependence on subsistence labour, with rapidly increasing material
productive power as the result of automation, thus increasing the
amount of free time one can use. There will also be an expanded
ability to solve problems and pursue new possibilities, and then to
bring such possibilities into reality; that is to say, it will expand one's
ability for futurization.

The development of this information productive power will offer
the individual more independence than can be enjoyed now.

Another point to be noted is the autonomous expansion of
creativity that will follow. The keynote of utopian societies in the past
has been the establishment of communal life through the common
ownership of the means of production, based more or less on the
prototype of primitive communism. This type of society has inevitably
operated with a relatively low level of productive power; but the future
information society will ensure more active voluntary communities,
because humans will be liberated from dependence on subsistence
labour, and because of the expanded possibilities for future time-
value realization.

As a consequence, utopian societies will move on from being
merely cooperative societies, where most time must still be given to
sustaining existence, to become dynamic and creative voluntary
communities. It is people with common goals who will form the new
voluntary communities, communities that will always be carried on by
voluntary activity and the creative participation of individuals;
individual futurization and group futurization will be harmoniously
coordinated with societal futurization. In the mature information
society of the future, nature communities, non-smoking communities,
energy conservation communities, and many other new types of
voluntary communities will prosper side by side.

Interdependent Synergistic Societies

My fourth vision is *the realization of interdependent synergistic societies*. A synergistic society is one that develops as individuals and groups cooperate in complementary efforts to achieve the common goals set by the society as a whole. The functioning societal principle is *synergism*, a new principle to replace the free competition of the current capitalistic society.

In the future information society, information utilities, whose structure of production is characterized by self-multiplication and synergy, will take the place of the present large factories, and become the societal symbol of the information society. These information utilities will be the centres of productive power, yielding time-value that will be the common goal of voluntary communities, because of *the self-multiplication* that characterized production in the information utility. Unlike material goods, information does not disappear by being consumed, and even more important, the value of information can be amplified indefinitely by constant additions of new information to the existing information. People will thus continue to utilize information which they and others have created even after it has been used, and, at macro level, the most effective way to increase the production and utilization of information will be for *people to work together to make and share societal information*. This economic rationality means that the information utility itself will become part of the infrastructure. It will be the force behind the productive power which gives birth to socio-economic values, and corresponding new socio-economic laws and systems will come into being as a matter of course. *Synergistic feedforward* will function as the new societal principle to establish and develop social order, with the resulting societies becoming voluntary communities.

Functional Societies Free of Overruling Power

My fifth vision is of *the realization of functional societies free of overruling power*. The history of the rule of man over man is long, continuing right into the present, simply changing form from absolute domination by an aristocracy linked with religion in feudal society to economic domination of enterprises in capitalist society, and to political domination by the bureaucracy in both socialist and capitalist society. The future information society, however, will become a *classless society*, free of overruling power, the core of society being voluntary communities. This will begin as informational and local communities comprising a limited number of people and steadily develop and expand. A voluntary community is a society in which the independence of the individual harmonizes with the order of the group, and the social structure is a multi-centred structure characterized by mutual cohesion. By 'multi-centred' I mean that *every individual and group in a voluntary community is independent, and becomes a centre.* 'Mutual cohesion' means that *both individuals and groups that constitute the centres share a mutual attraction to form a social group.* Behind this mutual attraction lies the common goal, the spirit of synergy, with the ethics of self-imposed restraints. In other words, as individuals pursue their own time-value, they work synergetically as a group to achieve a shared goal, and all exercise self-restraint so that there will be no interference with the social activities of others. This social structure is the overall control system of a voluntary community. In the political system, democracy based on participation of the citizens will be the general mode of policy making, rather than the indirect democracy of the parliamentary system. The technological base to support this participatory democracy will consist of (1) information networks made possible by the development of computer-communications technology, (2) simulation of policy models, and (3) feedback loops of individual opinions; with the result that policy making will change from policy making based on majority versus minority rule to policy making based on the balance of gain and loss to individuals in the spectrum of their areas of concern, both in the present and in future time. In policy making by this means, the feedback and accumulation of opinions will be repeated many times until agreement is reached, to ensure the impartial balance of merits and demerits of the policy decision as it affects individuals and groups with conflicting interests.

The present bureaucratic administrative organization will be converted into *a voluntary management system of the citizens.* Only a small staff of specialists will be needed to carry out administrative duties, officers who are really professionals responsible for the administrative functions. The bureaucratic organization of a privileged class will disappear. In this voluntary civil society, ruling, coercion and control over others will cease. Society will be synergistically functional, the ideal form that the information society should take.

Computopia: can it become a Reality?

Can these visions of Computopia be turned into reality? We cannot escape the need to choose, before it is established, either 'Computopia' or an 'Automated State'. These inescapable alternatives present two sharply contrasting bright and dark pictures of the future information society. If we choose the former, the door to a society filled with boundless possibilities will open; but if the latter, our future society will become a horrible and forbidding age.

As far as present indications go, we can say that there is *a considerable danger that we may move toward a controlled society*. This is seen in the following tendencies.

During the first fifteen to twenty years of their availability, computers were used mainly by the military and other government organizations and large private institutions. Medium and small enterprises and individuals were generally barred from using computer-communications technology, since large scale computers at the early stages of automation were extremely costly. This situation caused a significant delay in democratic applications of computers. Initially, computers were used mainly for automatic control and labour-saving purposes, rather than 'problem solving' applications. The development from automatic control of separate systems to integrated real-time control systems covering broad areas is increasing the danger of a controlled society.

The utilization of computers for major scientific and technological applications, such as space development, has led us to neglect the need for coexistence with nature, while our impact on nature has grown immeasurably. The development of 'big' science and technology has operated in such a way as further to increase the imbalance between human and nature systems.

If computerization continues in this direction, the possibility of a controlled society increases alarmingly.

However, I believe and predict that *the catastrophic course to an 'Automated State' will be avoided*, and that our choice will be to follow the path to 'Computopia'. I give two logical reasons for my confidence.

The first theoretical basis is that *the computer as innovational technology is an ultimate science.* By 'ultimate science' I mean *a science that will bring immeasurable benefits to humanity if wisely used, but which would lead to destruction if used wrongly.* Nuclear energy, for example, can be an extremely useful source of energy, but it could kill the greater part of the human race in an instant. The computer may, in one sense, be more important, as an ultimate science, than atomic energy.

If computers were to be used exclusively for automation, a controlled society, the alienation of mankind and social decadence would become a reality. But if used fully for the creation of knowledge, a high mass knowledge creation society will emerge in which all people will feel their lives to be worth living. Further, an on-line, real-time system of computers connected to terminals with communication lines would turn society into a thoroughly managed society if utilized in a centralized way, but if their utilization is decentralized and open to all persons, it will lead to creation of a high mass knowledge creation society. Similarly, if data banks were to be utilized by a small group of people in power to serve their political purposes, a police state would ensue, but if used for health control and career development, every person can be saved from the sufferings of disease and be enabled to develop full potentialities, opening up new future opportunities and possibilities.

So it is *not the forecasting* of the state of a future information society, *but our own choice* that is decisive. There is only one choice for us – the road to Computopia. We cannot allow the computer, an ultimate science, to be used for the destruction of the spiritual life of mankind.

The second theoretical basis of my confidence is that *the information society will come about through a systematic, orderly transformation.* The information society will be such that information productive power will develop rapidly to replace material productive power, a development that will bring about a qualitative conceptual change in production, from production of material goods to the production of systems. By this I mean the production of far-reaching systems that include everything from production systems for material goods (such as automated factories), to social systems (wired cities, self-education systems), to political systems (direct citizen participation systems), and even to ecological systems.

Obviously, information productive power centring on the computer-communications network will be the powerful thrust to bring about

societal systems innovations. New social and economic systems will be created continuously, and society as a whole will undergo dynamic changes, not the drastic social changes of the past, typified by the power struggles of ruling classes, wars between nation states and the political revolutions of mass revolt. It will be achieved through *systematic, orderly transformation*. As old socio-economic systems gradually become ineffectual and unable to meet the needs of the times, they will atrophy, and new, responsive socio-economic systems will take their place, in the way that a metamorphosis takes place with an organism, the useless parts of the body atrophying and other parts developing in response to the new demands.

Moreover, this systematic transformation of the societal structure will be brought about by citizen action, *changing means- and goal-oriented modes of action into cause-and-effect modes of action*. I have pointed out that human modes of action will become goal-oriented in the information society. These modes of goal-oriented action will evolve to the point where they function as a goal principle, to become the principle of social action. When this happens, social action will be logical, means-oriented action for the pursuit of common goals. So we can replace the term, 'goal–means-oriented action', with the term, 'cause–effect relationship', following the idea of Max Weber. In the information society, the social actions of citizens in general will become goal–means relationships that operate as cause–effect relationships.

The Rebirth of Theological Synergism

The final goal of Computopia is the rebirth of theological synergism of man and the supreme being, or if one prefers it, the ultimate life force, expressions that have meaning both to those of religious faith and the irreligious. This can be called the ultimate goal of Computopia. The relation existing between man and nature was the beginning of civilization. For many thousands of years man was completely encompassed by the systems of nature, which he had either to obey or be destroyed by them. Five or six thousand years ago, man succeeded in harnessing these systems of nature in a limited way to increase agricultural production, and the first civilizations were built. This marked the beginning of man's conquest of nature. But with the industrial revolution the conquest of nature meant the

destruction of nature, and now nature's retaliation has begun, the sequel to man's relation with nature that turned into destruction.

Now, a new relationship is beginning. At last, man and nature have begun to act together in a new ecological sense, on a global scale, in a synergistic society. At the base of this conversion of human society into an ecological system is the awareness of the limitations of scientific technology. It means awareness that scientific technology is simply the application of scientific principles, and that these cannot be changed by man, nor can he create new principles to work and live by. It is also a new awareness of the commonality of man's destiny, in that there is no place where man can live except on this earth, which first gave him life; from this very awareness is emerging the idea of a synergistic society where man and nature must exist in true symbiosis.

This is the assertive, dynamic idea that *man can live and work together with nature, not by a spirit of resignation that says man can only live within the framework of natural systems*; but, not living in hostility to nature, man and nature will work together as one. Put another way, man approaches the universal supra life, with man and god acting as one.

God does not refer to a god in the remote heavens; it refers to nature with which we live our daily lives. The scientific laws that we have already identified and are aware of are simply manifestations of the activity of this supreme power. The ultimate ideal of the global futurization society will be for man's actions to be in harmony with nature in building a synergistic world.

This synergism is a modern rebirth of the theological synergism which teaches that '*spiritual rebirth depends upon the cooperation of the will of man and the grace of God*', however it may be expressed. It aims to build an earthly, not a heavenly, synergistic society of god and man.

When we open the book of history, we see that when man brought about the accumulation of wealth and an increase in productive power, various choices had to be made. The Greeks built magnificent temples to Apollo and carved beautiful statues of Venus. The Egyptians built gigantic pyramids for their Pharaohs, and the Romans turned the brutalities of the Colosseum into a religious rite. The Chinese built the Great Wall to keep out the barbarians. Now man has made the fires of heaven his own, and left footprints on the craters of the moon.

We are moving toward the twenty-first century with the very great goal of building a Computopia on earth, the historical monument of

which will be only several chips one inch square in a small box. But that box will store many historical records, including the record of how four billion world citizens overcame the energy crisis and the population explosion; achieved the abolition of nuclear weapons and complete disarmament; conquered illiteracy; and created a rich symbiosis of god and man without the compulsion of power or law, but by the voluntary cooperation of the citizens to put into practice their common global aims.

Accordingly, the civilization to be built as we approach the twenty-first century will not be a material civilization symbolized by huge constructions, but will be virtually *an invisible civilization*. Precisely, it should be called an 'information civilization'. *Homo sapiens*, who stood at the dawn of the first material civilization at the end of the last glacial age, is now, after ten thousand years, standing at the threshold of the second – the information civilization.

Notes

1 Y. Masuda, *Computopia*, Tokyo, Diamond, 1966.
2 A. Smith, 'An Early Draft of the Wealth of Nations', in W. R. Scott, *Adam Smith as Student and Professor*, Glasgow, 1937.
3 W. W. Rostow, *The Stage of Economic Growth*, London, Cambridge University Press, 1960.

14

The Genesis of *Homo Intelligens*

This concluding chapter is written in accordance with my historical hypothesis that the information revolution will bring about not only the transformation of human society from *industrial society* to *information society*, but also the transformation of the species of man from *Homo sapiens* to *Homo intelligens*.

My vision of the new man is not as a spaceman of science fiction; my hypothesis is rather an attempt to anticipate the genesis of new man scientifically, by linking recent results from palaeoanthropology and sociobiology with advanced information science, and by use of historical analogy.

I suggest for the new man the name of *Homo intelligens*, as distinct from modern man, whom we call *Homo sapiens*. I consider that the genesis of *Homo intelligens* stems from the Creator's providence directed against modern man, who has brought upon himself the crisis of his own existence as a species.

The first and principle ground for my argument is a recent theory on the genesis of modern man as a species, based on results of research into human palaeoanthropology.

The theory so far generally accepted on the genesis of modern man may be summed up as follows: man's ancestor gave up living in the trees for some reason and began to walk upright; his fingers developed and he began to make tools; his brain was relatively bigger than those of other primates, such as anthropoids. These three factors – (1) upright walk, (2) invention of tools and (3) development of the brain – were regarded as decisive.

Recent remarkable developments in human palaeoanthropology have completely exploded this theory. A revolutionary and fascinating new theory is explained by the US anthropologist, Jeffery Goodman.[1]

He states that before the genesis of modern man there had already existed *Australopithecus*, *Homo erectus* and Neanderthal man as human species whose outstanding features were (1) upright walk, (2) invention of tools and (3) a developed brain, and each of these types of human coexisted with the other types of human species for a considerable period.

Australopithecus africanus is the species most often considered the direct ancestor of the human line . . . The australopithecines evolved manlike dentition, an upright walk, and a relatively larger brain . . . Lacking imposing size, strength, or vicious canines, *africanus* roamed the grassy savannahs of Africa in family groups gathering plant foods, scavenging at lion kills, and occasionally preying upon the less formidable fauna such as rabbits and turtles. It is not clear whether *africanus* had crude stone tools.

Homo erectus is the first hominid advanced enough to be classified in man's own genus . . . *Homo erectus* stood fully upright on a sturdy and muscular five-foot frame . . . *Erectus*'s average brain size of 1,000 cubic centimeters (ranging from 750 to 1250 cubic centimeters), two-thirds of that of modern man . . . Equipped with a tool kit including hand axes, he seems to have been a much more able hunter than *Australopithecus*.

The rugged Neanderthal skeleton reflects great strength; they were stoutly built and heavily muscled . . . The enlarged brain represents a clear advancement over *Homo erectus*; in fact, the average brain volume for some Neanderthal populations is slightly greater than that of modern man (1,360 cubic centimeters average). The Neanderthal's increased brain volume was probably devoted to control of his more massive and complex musculature. Their tool kit showed relatively little change or innovation . . . Richard Klein of the University of Chicago bluntly notes that they were 'rotten hunters'.

Modern man made his debut about 35,000 years ago when the Neanderthals were still roaming about in Europe, and came to replace the latter. What is worthy of special attention here is that modern man was basically different as a species from *Australopithecus*, *Homo erectus* and Neanderthal, about which Jeffery Goodman says,

The Neanderthals' retention of *erectus*'s large jaw and teeth also made them distinct from modern man. Another crucial difference is the construction and capabilities of the vocal tract . . . The Neanderthal vocal tract had only one effective sound chamber, the mouth, while the modern man has two, the mouth and pharynx, and can use rapid tongue movements to vary the size and shape of both passages to produce a wide range of sounds. Computer directed tests [by Dr Philip Lieberman, co-worker at MIT] showed that neither hominid could produce the variety of sounds essential to modern speech and language. As with present-day chimpanzees, the vowels a, i and u were impossible for *Homo erectus* and Neanderthal man.

Instead of the same old tool types made in the same limited number of ways, the tool kit of Cro-Magnon man [modern man] is an astounding advance over those of his predecessors in utility, variety, and complexity . . . Interestingly, the area of the brain that governs the fine actions of the hand required for advanced toolmaking and art lies very close to the area of the brain that controls the muscular movements required for speech.

While modern man's skull on the average is not particularly larger than Neanderthal man's, it has undergone great reorganization. Its new and distinctive high-foreheaded shape packages an even more radical evolutionary departure: the expanded frontal section of the brain, which controls nearly every distinctively human activity . . . While *Homo sapiens'* unprecedented brain structure, with its accentuated frontal lobe, brought with it qualitative changes astonishing in terms of the quantitative changes involved. The unique frontal section of the human brain (including the neocortex and cerebrum) makes possible man's greatly improved intellect, fine motor skills such as finger dexterity, and linguistic skills such as speech; it is also involved in individual behavior in social relationships, controlling traits such as mood, drive inhibition, and ethical judgment . . . Without such personality changes and detailed physical skills the emergence of modern culture would have been impossible.

Significance of the New Theory on the Genesis of Modern Man

The new theory on the genesis of modern man, which has disproved the established evolutionary theory at its foundation, provides a significant biological ground for rethinking the nature of modern man as a species.

The first point is that modern man as a species had features distinct from preceding human species, and that this was a decisive factor. More concretely, modern man had features such as the *developed frontal lobe, a complex vocal organ and outstanding finger dexterity.* In other words, we of modern man are a clear departure on the intellectual and cultural levels from Neanderthal man and the other preceding human species, notwithstanding our racial and individual differences. In plain words, even mentally retarded children and native peoples are equipped with intellectual and cultural capacities far superior to those of the early human species.

The second point is that, of the three characteristic features which made possible the genesis of modern man as a species, the development of the frontal lobe played a core role. If the frontal lobe had not developed, it would have been impossible to attain linguistic skills even if there were a complex vocal organ, because the development of language is impossible without a repetitive, intellectual process of thinking by which man memorizes and conceptualizes external objects or experiences and expresses them in particular symbols. Similarly, outstanding finger dexterity and the making of precision tools is at once a result of improving primitive tools by learning, by trial and error and through inventive ideas, a process which continued over a long period through many generations.

With the development of language, the invention of well-designed tools and the acquisition of high manual skills resulting from the use of the vocal organ and fingers, man accumulated new information and knowledge and stored it in the frontal lobe, which in turn developed the functions of the frontal lobe even further; the quantitatively and qualitatively developed frontal lobe in turn raised the content of language and tools. In this way these three intellectual organs as a whole in a spiral ascent expanded the intellectual and cultural capacities of modern man.

Emergence of Computers, New Communications Media and Robots

The second strong scientific basis of my argument on the genesis of new man is the emergence of *computers*, *new communications media* and *robots*, resulting from the remarkable development of advanced information sciences consisting of computer-communications science and knowledge engineering.

If, as mentioned above, the development of the frontal lobe, the complex vocal organ and outstanding finger dexterity are the critical elements in the genesis of modern man as a species, a correspondence can be seen between these features and the development of the computer, new communications media and robots. It is from this standpoint that these three offer strong scientific grounds for the possibility of the genesis of a new type of man, superseding modern man.

Computer-production of Intelligence Information and Knowledge

The computer is the third societal transformational means of production that man has developed since the beginning of recorded history. The first was the stone axe as the first universal tool; the second was the steam engine, the motive force of the industrial revolution of the eighteenth century. Unlike the first and the second, the computer is an epochal means of knowledge production, in that it produces large quantities of new information, not material goods.

In the past, mankind has invented four different kinds of epochal information technology – language, characters, printing and communication. These concerned expression and copying in the form of sounds and symbols of information created by the human brain, the original information being produced by the brain itself. However, the computer, which performs the three functions of memory, arithmetic operation and control, can produce large quantities of original information if the data and programs are given by man.

The progress of computer technology is quite astounding; the first generation computer (using vacuum tubes), which was developed in 1946 for the purpose of calculating the ballistics of intercontinental missiles, is now at the stage of the fourth generation (optimum devices), passing through the second generation (transistors) and the third generation (VLSIs).

The fifth generation computer, which Japan is now developing in international cooperation under a ten-year project which began in 1983, is planned as a computer capable of inferential deduction, association and learning like a human brain, instead of merely sequential data processing such as in the computers of today.

For this, R&D work is being done on three basic technologies. One concerns the development of a logic-type programming language, a programming language by which the computer is instructed how to process a given problem logically. The second is technology for the development of a parallel inference machine, which is intended to be capable of performing inference by syllogism one million times a second simultaneously and in parallel. This is a system for classifying inference into problem solving inference and objective-accomplishing inference, structuring a knowledge base and controlling it effectively.

Already a biocomputer has become a subject of R&D as the sixth generation computer. Recent results of neurobiology have made it clear that memory is established as bits of information are stored in a network of neurons connected with synapses, and also that the brain is equipped with a self-organizing function – an automatic programming function, in computer terms – of recognizing changes in the external world and of enabling proper action to be taken.

Furthermore, structure of the sodium channels that govern the information transmission of nerve cells has been clarified at molecular level, and it is expected that this knowledge will lead to a breakthrough in the development of a new information processing device, incorporating the functions of nerves.

The remarkable development of bioengineering will unravel one by one the mysteries of the complex mechanisms of information processing and transmission, and knowledge formation by the brain, and as a result, a biocomputer will some day be developed as the sixth generation computer equipped with structure and functions similar to those of the human brain.

The emergence of a computer as an artificial brain, at the apex of which will be the biocomputer, will greatly amplify the function of the frontal lobe. In the first place, the computer will substantially raise the quality of information and knowledge processed by the frontal lobe, because information created by the computer is highly intellectual and complex.

Before anything else, it is logically processed cognitive information that will serve man in the selection of optimum action. Furthermore, it is systems-oriented information which will combine intellectual information of different qualities, and at the same time is prognostic information that will offer models of complex social–economic phenomena and make predictions on future overall figures. Moreover, the computer will vastly expand the boundaries of space and time in relation to information and knowledge processed by the frontal lobe, because the computer, connected with a communications satellite, can process and communicate global information and knowledge at electronic speed.

The message from an astronaut, 'The earth is blue', expanded the spatial boundaries of our information and knowledge to the level of space at one stroke, and, needless to say, the computer provided the basic technology that made possible the launching of a manned artificial satellite into space.

Further, the Club of Rome's warning on 'the limits to growth' gave a deep, intellectual shock to the people of the world as a criticism of modern scientific civilization, and the figures on which the warning was based were calculated by the method of system dynamics, utilizing a large capacity computer.

And again, it was established that the destruction of tropical rain forests is the basic cause of abnormal weather worldwide, which in turn is responsible for the rapid growth of desert areas in the world. It was made clear from the analysis of tremendous quantities of data obtained from meteorological satellites how the destruction of tropical rain forests has disrupted the meteorological and biological cycle of the earth as a whole, a cycle by which solar heat is stored in the humid tropical rain forests and by which the stored heat is circulated by rising air currents to warm the temperate region far from the tropical areas.

Information and knowledge of this kind are entirely different from the type of information and knowledge so far obtained. Computer and information technology has made it possible for man to grasp complex phenomena in a many-sided structural way, and to analyse phenomena in their relations to the whole and from a long-term and circuitous route, far beyond the bounds of the capacity of man's frontal lobe. In addition, this technology will change man's criteria of values from narrow, direct and selfish interests to ethical criteria, oriented towards the interests of the whole.

This change in man's values will have an impact so strong that it will radically change our conception of the world, our thinking and our patterns of behaviour.

I call this kind of information and knowledge *intelligence information and knowledge*, distinguishing it from ordinary information and knowledge. If computers learn, they will have thousands and tens of thousands of knowledge bases; and if the work of a biocomputer with the same information processing capacity as the frontal lobe is integrated with that of the frontal lobe itself, information and knowledge of this kind will multiply explosively.

New Media – Formation of Autonomous Information Networks

Let us now proceed to the new media. More correctly, we should call the new media 'new information communications media', i.e. many kinds of advanced information/communications media of a new type made possible by the development of microelectronics.

In concrete terms, advanced information media include word

processors, intelligent terminals, facsimile systems, video cassette recorders and new-type communications media represented by digital communication systems, electronic exchange, optical communication systems and communication satellites. The application systems for these include cable TV, videotex, electronic mail, TV conferencing, home banking. But if the utilization of new media is limited to this, they cannot provide the scientific foundation for the genesis of new man. The reason is that all these methods of utilization of new media are seen as but an extension of industrial society, which has developed on the basis of production and consumption of material goods. For instance, even if TV channels are increased to thirty or fifty by cable TV, or if one can do one's shopping while at home, it will bring no change to the quality of life.

Only when these new media give full play to their effectiveness by the working of the frontal lobe multipied by the computer, or more specifically, when the new media are made the means of communication for what I have called intelligence information or knowledge, and only when they provide the basis for the formation of citizen-level information networks, can they become the scientific basis for the genesis of new man.

First, the development of *universal image symbols* can be considered suitable for the former or for the new information medium. The special features of these symbols may be summed up as follows:

1 This new medium of information is designed to express composite intelligence information structurally as well as in time series, and to appeal not to reason alone but also to emotions.
2 In consideration of the optimum appeal to man's senses, the power of cognition and understanding, it is designed to be capable of communicating meaning and content without prior training or learning.
3 All forms of means of communication, films, moving pictures, photographs, symbols and marks, are utilized for this medium of communication, but the use of words and characters is minimized.
4 In addition to forms, other media of communication such as colour, light, sound and smell, are also used in combination.
5 These media of information permit modifications such as synthesis and analysis, and free development in time series.

Take, for instance, a videotape utilizing these international image symbols. Needless to say, documentary films, animation pictures, statistics and graphs, and all the other existing media will be edited, combined, enlarged, substituted or repeated to give them dynamic expression.

However, the use of words, characters and other sequential expressions will be limited as much as possible because such expressions would hinder analogue understanding of composite and structural problems. It goes without saying that words and characters will be translated into different languages by use of an automatic translating machine.

The historic significance of these international image symbols lies in helping the people of the world to deepen their mutual understanding of the nature and issues of complex problems that are global in scale, such as population explosion, environmental disruption or nuclear war, irrespective of differences of states, nations, ideologies and language barriers.

As for the latter or new media of communication, mention must be made of the formation of a *voluntary civil information network*, having the following characteristics:

1 An individual-based network of information autonomously structured by individual citizens.
2 Optical communication, electronic exchanges, communication satellites, personal computers and other systems will be utilized as the technical foundation for such a network.
3 Personal computers owned by individuals will be interconnected by communication lines, and individual and group mail boxes will be provided to permit mutual switching of messages. Further, information and data common to individuals and individual opinions and data for discussion will be stored in such boxes for groups.

As an example of an autonomous information network one can point to a citizens' autonomous no-smoking network. This is a network of no-smoking movements now developing throughout the world, characterized by its being formed by those who are carrying on practical no-smoking activities as follows:

1 A close network is formed by citizens who are bound by the common objective to quit smoking.
2 They are actually engaged in no-smoking activities.
3 They are developing activities to urge other people to quit smoking according to a common programme.

The historic significance of the formation of citizens' autonomous information networks lies in the fact that such networks closely bind together individuals who share a common problem awareness or a common objective, according to the principles of consensus and participation, and lead them to take common action for that common objective.

In this way, citizens participating in autonomous information networks will discuss problems concerning the accomplishment of common objectives, sum up their discussions, and decide on a final solution by participation of all members, each one taking autonomous action for the purpose in a way suitable to his or her own position. This is precisely what is meant by direct participant democracy.

Robots – Liberation from Productive Labour

Let us now discuss the question of the robot. We can define the robot as a 'machine that can autonomously perform physical and mental labour in the same way as a human being does'.

Technically speaking, according to Professor Yoji Umetani at the Tokyo Institute of Technology, such a robot consists of three sub-systems: (1) artificial intelligence (arithmetic calculation, memory, power of distinguishing, learning, inference, etc.); (2) sensor element (size, weight, temperature, sound, shape, colour, smell, taste, tactile pressure, etc.); and (3) effective mechanical element (fingers, hands, arms, feet, etc.).

At present, robots are entering the stage of autonomous-type robots, capable of sensory judgement with sensors (second generation robots) from the stage of automatic robots repeating simple operations. Representative of these robots is the unit designed to work under extreme conditions – a robot that operates autonomously under hazardous conditions, such as inspection and repair of nuclear reactors, deep sea exploration and rescue work in fires.

For instance, Professor Keinosuke Miwa of the Engineering Department of Waseda University and his group have succeeded in developing a super-small medical robot that creeps inside human internal organs and blood vessels to irradicate affected parts with a laser beam. This device is a length of optical fibre with a shape-memorizing piece of alloy (which returns to its original shape after cooling or heating) at the tip. The alloy metal piece contains an inlaid VLSI.

Another development is a cell-analysing robot, which not only examines the tissues and structures of cells by the use of an ultrasonic microscope, but can also observe the internal conditions of cells three-dimensionally, including sound, speed, viscosity and density. A further application is a robot designed to aid the blind, which, keeping step, moves one metre ahead of the blind person and warns of any deviation from the correct course, stops at an intersection, and warns of danger if there is an obstacle ahead.

With the development in the near future of the fifth generation computer or a biocomputer, we will move into an era of intelligent robots of the third generation, capable of inference and learning by themselves. In the twenty-first century robots with a wide variety of intelligence capability will find their way into offices and factories, and even into homes.

The basic function of a robot is to do physical and intellectual work in place of man. The replacement of human labour by robots is seen not only in welding, painting, assembling and handling work in factories, but is already occurring partially in offices, commercial establishments and the service industry. It is generally accepted by most robot engineers that in the near future most productive operations in factories will be done by robots.

The increasing application of robot work will make it necessary to face one of two alternatives – mass unemployment or liberation from productive labour. This will call for human wisdom to choose *liberation from productive labour* and reject mass unemployment.

In my view, the liberation of man from labour that can be done by robots will be achieved by the introduction of the following transformational social and labour systems:

1 First, employment opportunities will be increased and working hours shortened by work-sharing and dual job systems, while bold employment expansion measures such as substantially increased holidays and vacations will be necessary.
2 Second, in addition to employed labour, forms of work will be diversified, such as contract and independent labour (independent consultants and independent work teams).
3 Third, the path will be opened up for a wide variety of new professions to be created, including systems engineers, programmers and other professionals related to the information industry; bioengineers, researchers, engineers and operators related to optoelectronics and

high technology industries; editors, fashion designers and veteran engineers in their respective fields who participate as advisers in voluntary citizen bodies. Many thousands of new professions and occupations can be expected.

Possibly by the end of the first half of the twenty-first century we will be completely liberated from work for production, and necessary working hours will be greatly reduced. A four-day working week and a two-month annual vacation system, and even employment forms such as a system of six months work and six months free time will be widely established.

But many people take a serious view of the liberation of man from productive labour, and regard it as the loss of something worthwhile. So far, labour for producing material goods has been a source of satisfaction, an important social act that is a source of happiness. But it is argued that if robots deprive man of the satisfaction of producing material goods, in what will people find the joy of living and something worth working for?

On this, Professor Ezra J. Mishan at Victoria University of Canada states that, 'so far labor has been the source by which man is bound to reality in his life, and also the source of self-respect for man, and . . . this labor is already disintegrating'.[2] Further, Adam Schaff, director of the Europe Social Science Research Institute, states that, 'work in the traditional meaning of the word has been the most basic activity in our life and we are now going to deprive man of this very thing worth living for'.[3] Similarly, Pope Paul II, referring to the microelectronics revolution, is reported to have warned against the danger of this new revolution, especially to young people.

Contrary to these apprehensions, I am of the opinion that the liberation of man from productive labour will bring to man a new objective worth living for, in the form of the *creation of time-value*, to replace the joy of producing material goods. Here, by 'time' I do not mean time in terms of hours and minutes, but 'space time' with a tangible content, or time in terms of a constantly changing situation. By 'time-value' is meant 'value created by designing future time', and given a literary expression, it is 'value by realizing what is designed on an invisible canvas of future space'. More objectively, it is 'value created in the time process of transforming the present situation into a desirable situation'.

Let me give an example. If law student A, wishing to become a lawyer, passes the examination to enter a legal training and research institute and is qualified to practise law, this means a change in his situation from a student to the more desirable position of lawyer, the creation of time-value. From the *value theory of situational transformation*, the value of producing material goods is essentially the same process as the value of creating time. Thus, the process of material goods-producing value is the process of transforming a physical situation of natural raw material to a physical situation of useful goods, except that the process of creating time-value is a more complex situation-transforming process of a higher order than the process of creating material values.

It must be emphasized that if robots only liberate man from productive labour to provide more free time, this will do no more than create idle people like the free citizens of Rome and the knights of medieval Britain. The real historic significance of robots should be sought in the fact that according to the theory of situation-transforming value, mankind will move in a big way from the satisfaction of material needs to the realization of time-value, and that mankind will realize human value of a higher order worth living for.

Genesis of *Homo Intelligens*

The third essential ground for my argument on the genesis of new man is the theory of 'gene–culture coevolution'. About this, Charles J. Lumsden and Edward O. Wilson stated that:

Gene–culture coevolution is a complicated, fascinating interaction in which culture is generated and shaped by biological imperatives, while biological traits are simultaneously altered by genetic evolution in response to cultural innovation. We believe that gene–culture coevolution, alone and unaided has created man . . .

One conception of gene–culture coevolution can be summarized very briefly as follows: to start, the main postulate is that certain unique and remarkable properties of the human mind result in a tight linkage between genetic evolution and cultural history. The human genes affect the way that the mind is formed – which stimuli are perceived and which missed, how information is processed, the kinds of memories most easily recalled, the emotions they are most likely to evoke and so forth. The processes that create such effects are called the epigenetic rules. The rules are rooted in the particularities of human biology, and they influence the way culture is formed. For example, outbreeding is much more likely to occur than brother–sister incest because individuals raised closely together during the first six years of life are rarely interested in full sexual intercourse. Certain color vocabularies are more likely to be invented than others because of other sensory rules entailing the manner in which color is perceived . . .

The translation from mind to culture is half of gene–culture coevolution. The other half is the effect that culture has on the underlying genes. Certain epigenetic rules – that is, certain ways in which the mind develops or is more likely to develop – cause individuals to adopt cultural choices that enable them to survive and reproduce more successfully. Over many generations these rules, and genes prescribing them, tend to increase in the population. Hence, culture affects genetic evolution just as the genes affect cultural revolution.[4]

Let me explain this theory of gene–culture coevolution in more detail as I understand it. While the acts of ordinary animals are one-sidedly determined by the genes, man generates culture through the function of the brain and mental capacity. The characteristics of culture generated in this way are affected by the genes. However, as culture generated in this way develops, culture in turn comes to have an effect on genetic evolution. In this way, man's genes and culture follow the course of coevolution, mutually influencing each other.

There are two points in this theory of gene–culture coevolution that claim my special attention: the first is that the brain and mental faculty are basic to culture generated by man. This process is closely related to the frontal lobe and the complex vocal organs of present mankind. This leads me to predict a rapid development of a new culture by means of the computer and the new media man has acquired.

The second point is that when a new culture develops in the future, it will cause a radical change in the genes of present mankind, offering the possibility of the genesis of new man as a new species. On this point, Lumsden and Wilson say, 'we believe that gene–culture coevolution, alone and unaided, has created man [i.e., present man]'.[5]

If we boldly apply this theory of gene–culture coevolution, can we not project that a new man will emerge as a new species, with high intelligence information and knowledge such as man has never before acquired, made available by the computer and, as voluntary citizen information networks are formed by the new media, and, as man is liberated from productive labour, by robots? What I am referring to here is the genesis of the new man, *Homo intelligens*.

However, anatomically, *Homo intelligens* will have no new traits as a new species, because no biological changes will have taken place in the organs or living tissues of the new man. But if there are major changes in man's basic desires, values, way of life and patterns of behaviour, it will mean the generation of a new culture, and if this leads to a change in man's genes, we may be able to regard this as a biological change for a new species at genetic level.

If we are allowed to express this point as Lumsden and Wilson would, a fundamental change in man's culture is supposed to lead in due course to a genetic evolution in present mankind and a fundamental change in basic traits.

What then is the new man *Homo intelligens* as I understand him? First, new man is man of high intelligence as the name signifies. Modern man, *Homo sapiens*, acquired a surpassing intellectual capacity as a result of the function of the frontal lobe compared with more primitive human species, and came to rule the animal kingdom as a whole as the lord of creation. But the fundamental characteristic of the intellectual capacity of modern man is his cleverness. This cleverness is based on narrow, short-sighted information and knowledge, and he is selfish and oriented towards satisfaction of his material needs. His intellectual capacity is therefore characterized by cleverness and shrewdness. In this sense, the intellectual level of modern man cannot be said to be of a high order.

But it was cleverness which was necessary for modern man to survive, when weak and powerless, he was forced to leave his peaceful life in the forest where he lived on nuts and fruit and, due to some external factor, suddenly had to live on the savannah where there were mammoths and swift herbivorous animals with which he had to compete.

Modern man later developed new media of information, language, characters and printing as well as bows and arrows, hoes and ploughs and other tools one after another until an industrial society was built on the basis of scientific knowledge and powered machinery.

However, because the motive force for development and progress was crafty intellectual capacity rooted in selfishness and material wants, knowledge, techniques and social systems born of this were necessarily characterized by the two basic features of material production and control. It is because of this that knowledge and technology created by modern man is centred on the domination of nature (including plants and animals) and of man by man. In particular man has developed military knowledge and technology for aggression against other nations and states.

Human history has therefore been a bloody history of wars between power-hungry chieftains and states, plunder and conquests, a history of sharp market acquisitive competition among avaricious merchants and entrepreneurs.

Now, because of this very cleverness, modern man stands at the brink of self-destruction as a species, sitting on the peak of a material civilization he has built for himself:

- There are at present about 50,000 nuclear warheads stored by the United States, the USSR and other nuclear weapons states. If even 10,000 megatons of nuclear weapons (equivalent to 500,000 bombs of the Hiroshima type) were used in nuclear war, 1,150 million people would be killed instantly and 1,095 million others would suffer burns and injuries. Some 10 million people out of a billion survivors would suffer from leukaemia, and one-third of the babies born to survivors would have hereditary abnormalities (Potikov, Soviet Academician).
- There are potential factors for the recurrence of a world depression similar in scale to the Great Depression of the 1930s, not only in both industrialized and developing countries but also in the oil-producing states. Socialist countries would not be immune, and unemployment in the industrialized nations would reach 32 million (Schmidt, former West German Chancellor).
- The balance of medium-term and long-term debts of developing countries stood at $530 billion in 1982 and would reach $600 billion by 1983 (World Bank annual report, 1983).

These reports graphically present the fact that the nation states, and the economic and social systems modern man has energetically created with his cleverness, have failed to function properly, and as a result mankind as a whole is faced with the threat of total ruin.

In contrast, new man as *Homo intelligens* will build a civilization entirely different from that built by *Homo sapiens*.

The first and the most decisive action *Homo intelligens* will take will be to overcome mankind's crisis of existence. This will be carried out by horizontal social transformation of a peaceful revolutionary type in which individual citizens will participate, and not through social transformation from above of a violent revolutionary type by force of arms or power such as man has so far used.

When mankind's crisis deepens further, citizens aware of their mission to save mankind will arise in tens of thousands of areas. Such citizens will be ordinary people – dentists, housewives, retired people, students and people in all walks of life – and will not be saviours like Buddha or Christ, nor dictators like Napoleon or Hitler. These people will rally other people around themselves, and these autonomous networks of people will expand and multiply irrespective of occupation, sex or age, networks that will grow on a worldwide scale irrespective of national boundaries and racial distinctions.

The formation of citizens' autonomous networks determined to save mankind from crisis is similar to the formation of a social amoeba as a biological phenomenon.[6]

Dictyostelium discoideum – a typical species of cellular slime mould amoeba – repeats cell division and multiplies while devouring bacilli around it (growth period), and when it has eaten all the bacilli around it, it gradually reassembles around one of them to form a slug-like body (assembling period) and begins to move (moving period). After completing its movement, it is formed into a fruit body (hymenium) consisting of spores and a handle (hymenium formation period). The spores germinate to become a cellular slime mould amoeba, which starts cell division again (germination period). Because of this behaviour, Professor Bonnar of Princeton University calls cellular slime mould amoeba 'social amoeba'.

Man as a species is faced with a crisis of existence, and if my bold hypothesis is applicable, modern man is at the end of the growth period as in the case of the amoeba, and is about to enter the period of assembling. If I am permitted to express this in Professor Bonnar's terms, it may be said that new man will go through a sociobiological change to social man.

When an amoeba, the simplest unicellular monad, is endowed by the Supreme Being with this intellectual instinct to overcome the extinction of its species, how can one reject as an overly optimistic fiction the concept that *Homo intelligens* equipped with highly sophisticated biological intellect will be changed to *social man* who will launch a global action when faced with a crisis of mankind as a species?

Notes

1 Jeffery Goodman, *Genesis Mystery – Explaining the Sudden Appearance of Modern Man*, New York, Times Books, 1983.

2 Ezra J. Mishan and Tadanao Kurakake, 'Growth, Government and Science and Technology', Tokyo, *Monthly NIRA*, No. 10, 1982.

3 Adam Schaff, Speech at Club of Rome Assembly, Tokyo, 1982.

4 Charles J. Lumsden and Edward O. Wilson, *Promethean Fire: Reflections on the Origin of Mind*, Cambridge, Mass., Harvard University Press, 1983.

5 Ibid.

6 Jiro Ota, *Amoeba*, Tokyo, Japan Broadcasting Publishing Association, 1977.

Glossary

affective information   information that is based on sensitivity and production of emotion. It embraces all the information that conveys sensory feelings, such as 'comfort', 'pain' and the emotional feelings of 'happy' and 'sad'.

arts industries   the industries that process, retrieve and service affective information or produce and sell related equipment. *See* **affective information, information industries.**

balanced feedforward   the subject of action and the external environment exercise mutual control if their relations are in balance. *See* **feedforward.**

CAI   computer aided instruction, computer-oriented self-learning system.

cognitive information   information that is a projection of the future; it is logical and action-selective. The projection means that the cognitive information is used for detecting and forecasting.

Computopia   computer'utopia, an ideal global society in which multi-centred, multi-layered voluntary communities of citizens participating voluntarily in shared goals and ideas flourish simultaneously throughout the world. *See* **voluntary community.**

controlled feedforward   the subject of action controls the external environment in the feedforward process. *See* **feedforward.**

dependent feedforward   the subject of action is dependent on the external environment in the feedforward process. *See* **feedforward.**

environmental information   information concerned with relation between an organism and the external world in order to maintain its existence. *See* **organismic information.**

ethics industries   the industries that process, retrieve and service ethical information.

feedforward   control in moving toward a goal, and viewed from the standpoint of the subject of action, it means a controlled development of the current situation to change it to a more desirable situation. *See* **dependent feedforward, controlled feedforward, balanced feedforward, synergistic feedforward.**

field the space with concrete content within which the subject of action acts with conscious purpose. *See* **process, subject of action.**

futurization future actualization; this implies actualizing the future, bringing it into reality. Expressed metaphorically, it means to paint a design on the invisible canvas of the future, and then to actualize the design.

GIU global information utility, a global information infrastructure using a combination of computers, communication networks and satellites. *See* **information utility.**

historical analogy societal foreseeing approach based on historical hypothesis; the past developmental pattern of human society can be used as a historical analogical model for future society.

information an informed situational relation between a subject and an object that makes possible the action selection by which the subject itself can achieve some sort of use-value. *See* **affective information, cognitive information.**

information cycle informative cycle of subject–object–signal–action. The subject receives a signal from the object, identifies the signal and evaluates it according to an acquired standard of judgement, selects a course of action, and finally achieves some use-value by implementing the action.

information epoch the span of time during which there is an innovation in information technology that becomes the latent power of societal transformation that can bring about an expansion in the quantity and quality of information and a large-scale increase in the stock of information.

information gap the relative absence of information processing and transmission technology between industrialized and developing countries, to which must be added the human factors of levels of intellectual development and behavioural patterns in such countries.

IIO International Information Organization.

information industries the industries that process, retrieve and service cognitive information, or produce and sell related equipment. *See* **knowledge industries.**

information society a society that grows and develops around information, and brings about a general flourishing state of human intellectual creativity, instead of affluent material consumption.

information space the field provided within the new space, which is connected with the networks of information, characterized by two features: (1) it does not have boundaries like a territorial field and (2) in this field elements related by objective-oriented action are related to each other through information networks.

information utility   an information infrastructure consisting of public information processing and service facilities that combine computer and communication networks. From these facilities anyone anywhere at any time will be able easily, quickly and inexpensively to get any information which one wants to get.

informational voluntary communities   a completely new type of voluntary community. It is the technological base of computer-communications networks that will make this possible. *See* **voluntary community.**

knowledge   cognitive information that has been generalized and abstracted from an understanding of the cause-and-effect relations of a particular phenomenon occurring in the external environment. *See* **cognitive information, information, technology.**

knowledge industries   the industries that produce, sell service knowledge and knowledge related equipment. *See* **information industries.**

objectification of information   the separation of information from its subject.

opportunity development   research and development of possibilities of future time usage or creating new values in rapidly changing environmental conditions.

organismic information   information concerned with physiological functions to maintain life within the body of the organism. *See* **environmental information.**

participatory democracy   a form of government in which policy decisions both for the state and for local self-government bodies will be made through the participation of ordinary citizens.

problem solving   a method or means of eliminating risks that may stand in the way of accomplishing an aim.

process   the development in time of a situation created artificially by the interaction between the purposeful action on the field of the subject of action and the reaction of the field to it. *See* **field, subject of action.**

quaternary industries   a new classification of industries; it is reasonable to distinguish information-related industries from service industries, and classify them as quaternary industries to provide a clear concept of the industrial structure of an information society.

self-multiplication of information   the quality of information is raised by adding new information to what has already been accumulated, and the accumulation of information leads to further accumulation of information which in turn means still further accumulation of information through time and space.

situational reform   a process in which the current situation is changed into a new situation that is consistent with the subject's goal, as expressed by the following formula:

Situation A $\rightarrow$ Situation A' $\rightarrow$ Realization of a value
(current situation)   (desirable situation)   (satisfaction of a want)

See **situational relation.**

situational relation   the relation between a subject and an object that comes into being in a particular situation, and to which there are three conditions: (1) there must be a subject and an object (in this case the environment surrounding the subject), (2) the subject must receive impulses from the object, and (3) the subject takes action in response to these impulses. *See* **situational reform.**

societal technology   this technology has four fundamental characteristics: (1) many different kinds of innovational technology are joined together to constitute one complex system of technology; (2) these integrated systems of technology spread throughout society as a whole and gradually take root; (3) as a result, there occurs a rapid expansion of a new type of productivity; (4) the development of this new type of productivity has a societal impact great enough to bring about the transformation from traditional society to a new society.

subject of action   the subject which works on the field with objective-consciousness. Such may be any individual, group of individuals, or organizations engaging in social action with deliberate purpose. *See* **field, process.**

synergy   a combined functional action by a group to achieve a common goal.

synergistic feedforward   the subject of action and the external environment are in mutually complementary relations and act together to reach a common objective.

systems industries   the structure of the systems industries will consist of a complex of industries formed by linking up existing industries with the information industries.

technology   cognitive information that is useful in effectively carrying out production-oriented labour requiring a certain degree of prescribed expertise. *See* **cognitive information, knowledge.**

theological synergism   the assertive, dynamic idea that man can live and work together with nature, not by a spirit of resignation that says man can only live within the framework of natural systems; but, not living in hostility to nature, man and nature will work together as one. Put another way, man approaches the universal supra-life, with man and god acting as one.

time-value   the value which man creates in the purposeful use of future time. Put in more picturesque terms, man designs a goal on the invisible canvas of the future, and goes on to attain it.

time-sharing system   the system by which several users have access to a computer simultaneously.

TSS   *See* **time-sharing system.**

voluntary community   a community in which a group of people carry on life together voluntarily with a common social solidarity. The fundamental bond to bring and bind people together will be their common philosophy and goals in day to day life. *See* **informational voluntary communities.**

Index

Page numbers in *italics* indicate tables and figures.

Index compiled by Geraldine Beare